SCIENCE, POLITICS AND THE COLD WAR

GRETA JONES

ROUTLEDGE
London and New York

First published in 1988 by
Routledge
11 New Fetter Lane, London EC4P 4EE

Published in the USA by
Routledge
in association with Routledge, Chapman & Hall, Inc.
29 West 35th Street, New York NY 10001

© 1988 Greta Jones

Printed in Great Britain

All rights reserved. No part of this book may be reprinted or reproduced or utilised in any form or by any electronic, mechanical or other means, now known or hereafter invented, including photocopying and recording, or in any information storage or retrieval system, without permission in writing from the publishers.

British Library Cataloguing in Publication Data

Jones, Greta
   Science, politics and the Cold War.
   1. World politics — 1945- 2. Science —
   Political aspects — History — 20th
   century
   I. Title
   327',09171'3   D842
ISBN 0-415-00356-3

Library of Congress Cataloging in Publication Data
ISBN 0-415-00356-3

Contents

Acknowledgements
Introduction
1. Between the Wars   1
2. Proletarian and Bourgeois Science   16
3. Into Two Camps   38
4. Race: a New Beginning   59
5. The Arms Race and the Scientists   79
6. Towards a Nuclear-free World   96
7. The New Right   119
8. Conclusion   136
Bibliography   141
Index   145

Acknowledgements

Material for this book has been collected over a number of years. I would like to thank the archivists and librarians of the following institutions for their help during that period: University of Ulster; Queen's University, Belfast; Bodleian Library, Oxford; University of Warwick; Imperial College, London; University College, London; Kings College, London; the British Library; the Royal Society; University of Liverpool; Nuffield College, Oxford; the Public Record Office, Kew; Colindale Newspaper Library, London.

I would also like to thank Professor Lawrence Badash and Sir Rudolf Peierls for their assistance. The opinions expressed in this book and any remaining errors are, however, my own.

Introduction

Most people agree about the importance of science. Its effects are seen everywhere. But not all realise the close relationship which exists between political struggles and science. Many historians have treated science as an established fact. It is there but it is only peripheral to political history. However, in an era in which so much state planning, so many techniques and practices of social organisation or political control are based or claim to be based upon science, it can never be treated simply as a given. The neglect is compounded by another factor. Whilst we are aware of the ways in which literary and artistic culture are shaped by politics and vice versa, and we accept that, for example, an intellectual history of the 1930s should include this, less attention has been given to the role of science in culture.

This book argues that science was central to several crucial political battles of the post-war world. Most scientists, it must be admitted, were not, at least willingly, politically engaged, but this is also true of the majority of those engaged in other intellectual work. Moreover, scientists were often involved in politics through non-scientific as well as scientific organisations and they aimed to influence rather than to direct or lead. Nonetheless, to understand the culture of the Cold War, a reassessment of the role of science is needed.

In 1945, the war in Europe came to an end. Gradually over the following years the alliance between Britain, the United States and the Soviet Union was disrupted by mutual suspicion and hostility. By 1948, the Cold War rather than war-time co-operation had come to dominate politics.

This was, to many on the left at the time, a tragic failure. They had hoped to extend the alliance which had defeated Hitler into the peace. The Second World War had brought much hardship and tragedy but, on the positive side, it had led to the defeat of Fascism. This defeat had been organised around an ideology of liberal democratic idealism. At times, for considerable numbers of people this had taken on a socialist complexion, which had been manifested in the new friendliness to the achievements of the Soviet Union, replacing the largely indifferent or hostile attitude of the inter-war years. Anti-Fascist ideology emphasised equality and human rights. In Britain, at least, there had arisen a new willingness to use the state for economic planning and, perhaps more significantly, there had been increasing emphasis upon social welfare and a better post-war deal for the poor. But in the aftermath of the Second World War and the disintegration of the optimism of the war-time alliance, how far would these ideals survive?

One section of the community, in particular, felt the increasing strain of events after 1945. Though they have been somewhat eclipsed in the popular mind by the literary figures of the 1930s, that decade had seen the emergence of a group of left-wing scientists committed to substantial social change. Some were liberals, a proportion of them committed socialists, and their achievements have not always received due recognition. The Second World War was for many of these a unique opportunity to participate in fighting Fascism and to help influence the post-war future.

These scientists, however, found themselves in 1945 in an invidious position. The development of atomic power, in particular the atomic bomb, put them at the forefront of the political struggles of the late 1940s. This led to two things: increasing surveillance of, and political pressure upon, scientists and, in the case of both the USSR and the West, an attempt to redefine the ideology of science to fit the competing political and social systems of the super-powers.

A political and ideological struggle took place. The intellectual right in science revived, became more self-confident and attempted to reconstitute science in a more acceptable political form. This book is about that struggle in science between right and left over the content and meaning of science and its political direction.

Atomic power was, for obvious reasons, one of the most crucial areas of struggle. On the whole, very few voices were raised in this period against the use of atomic power for industrial purposes. In fact, the opinion of most scientists was that, in the post-war reconstruction of Europe, atomic power would of necessity play an essential role. But many scientists did feel that there should be internationalisation of atomic energy and also of the atomic weaponry which the USA had developed in 1945 and which Britain was in the process of developing. One member at least of the British Cabinet shared this view in 1945. It rapidly became clear, however, that governments of both East and West took a dim view of scientists' involvement in political action in support of these goals. Yet, after initial setbacks, a concerted attempt was made through scientists' organisations to persuade first government, and then public opinion, of the desirability of an end to the atomic arms race.

Introduction

Scientists were in a unique position. A proportion were fully committed to the atomic weapons programme. But there were also those who were adamantly opposed and who were knowledgeable and well versed in the implications of atomic weaponry. By the mid 1950s their work had helped revive grass-roots opposition to the atomic arms programme in Britain.

Yet this achievement was bought at considerable cost. During the 1940s and 1950s, a cultural offensive in science came from the right, organised by groups such as the Society for Freedom in Science (SFS) and under the auspices of the Congresses for Cultural Freedom. These congresses became a feature of Europe in the 1950s. They were intended to emulate the Soviet practice of using conferences to influence the political climate among intellectuals. They were aimed at isolating the left and removing the liberal centre from their influence.

This was not unsuccessful and the success was not simply brought about by the pressure many scientists felt from government and the worsening political climate. In 1948 the prestige of left-wing science was dealt a serious blow when the USSR adopted Lysenkoist biology. The theories of the Soviet agronomist Lysenko ran counter to modern genetics as developed in Europe and the USA. The latter suggested that heredity was embodied in units of inheritance called genes. Although random mutation or variation occurred in these, they passed from generation to generation largely unchanged. Precise mathematical formulae could describe the process of inheritance. Lysenko, on the other hand, believed heredity could be changed more rapidly by the influence of the environment upon the organism. This process could be demonstrated by the process of vernalisation — the freezing of winter wheat seeds and their planting in the spring — in order to increase wheat yields.

With the help of some Soviet philosophers a highly elaborate attack on Mendelism as it was called — after Mendel the founder of modern genetical theory — was launched. This stigmatised it as bourgeois, idealist and undialectical. Suspicion in the Soviet Union of a link between conventional genetics and racism, the hope that Soviet agricultural production would be increased by Lysenkoism and the belief that socialism might lead to the development of distinctive sciences which better expressed socialist ideals led to Lysenko's victory in 1948.

Some of the left in the West blundered into Lysenkoism as a response to this change in the Soviet Union. The result was that they found themselves in a quandary. The Soviets were advocating sharp demarcations between science based on 'socialist' and 'capitalist' principles, just at the moment when the left in the West were struggling to maintain some form of common cause with the liberal centre. Even many socialist biologists were unhappy at what amounted to an attack on the normal practice of their science inspired by the dubious principles of Lysenkoist biology. Politically and tactically, leaving aside the question of the reliability of Lysenko's results or their philosophical implications, the Lysenko controversy was a blunder. It served to disillusion many liberals both with the USSR and with the political programme of left-wing scientists, and to isolate the latter.

Lysenkoism was pounced upon by the right. For them it was a perfect illustration of what they had been saying about socialism and science. In particular, their arguments about its authoritarianism, the unsoundness of its philosophy and above all, the incompatibility between true science and socialism appeared to be validated. Certain intellectuals went further. Throughout this period, the right produced exegesis about the close connection between liberal capitalism and the production of scientific knowledge. It was an exegesis which received a shock on the launching of the Soviet sputnik in 1957, but it was one which, none the less, had a powerful appeal. It linked a free market in scientific ideas and individual initiative to rapid technological progress. It contrasted this with the rigidities of state planning, the curtailment of freedom of research and the creation of bureaucratic monoliths to direct science, which it claimed was typical of socialism. Not all scientists were convinced by this argument, particularly in an age which paradoxically saw, in the West, increasing state support and intervention in science and growing political restrictions on scientists. None the less, the effect of the Cold War was to produce a coherent right-wing view of the role of science in society.

If many left scientists had an uneasy conscience about Lysenko, it had to be acknowledged that a problem existed in contemporary biology and genetics. When the revelations about Nazi eugenics in the death camps came to light in 1945, some scientists clearly felt that whereas Lysenkoism might be wrong, some answer had to be found to explain why so many German geneticists and doctors had been involved routinely in the practices brought to light in these camps. In other words, whilst Lysenkoism might be a wrong solution, the question, 'Was modern genetics racist in its character?' needed tackling.

Though the post-war world had brought increasing international tension, some of the ideals of the wartime struggle against Fascism had survived, first in a commitment to greater human equality of treatment as expressed in the United Nations' declaration on human rights and, second, in an organisation, the United Nations, pledged to international co-operation to promote these rights. On the question of race, the United Nations made efforts in the late 1940s and 1950s to confront racism in genetics and biology. As the United States, Britain and the USSR were at least formally committed to ideals of the United Nations Charter on human rights, considerable leeway existed for a propaganda effort on behalf of human equality. This was not altogether successful. Whilst progress was made, the climate of the times encouraged counter-attacks on the ideal of human equality. None the less, by the 1950s the proponents of racism in biology and genetics experienced isolation and a drawing back to the redoubt. The conservative ideology of the 1950s was not based on the pre-war Fascist or proto-Fascist ideology. It adopted American ideals of the free market, constitutional guarantees and free access to resources. In practice it also accepted a degree of state welfare. It was a conservatism which emphasised consensus rather than conflict or division. Without these characteristics, its appeal to the liberal centre might have been much more circumscribed. Yet by the late 1950s, this liberal democratic vision of the West was challenged. The maintenance of nuclear weapons meant more government secrecy and control, not less. There were no constitutional guarantees or free access in this area. By the mid 1950s, some scientists together with other intellectuals who had been successfully wooed by the SFS and the Congress of Cultural Freedom — or who had simply been alienated from politics in the post-war world — were seriously re-evaluating their position on the nuclear arms race. In addition, it was becoming clearer that the exigencies of the Cold War and the need for a cultural offensive against Communism had distracted attention from the issue of Nazism in biology and genetics.

In addition, the role of science in society and its relation to political structures still kept coming up. In the 1950s and 1960s, techniques and practices employing the language of scientific authority in any sphere from educational assessment to social work, proliferated. In the United States, the Camelot project for the scientific assessment of third-world insurgency in the 1960s brought scientists face to face with the question of the connection between science and politics.

By the 1960s there was increasing scepticism about the role

Introduction

science played in buttressing the existing power structures in society and an attempt, by some, to remould science to fit a better and different future. This was one reason why, by the early 1970s, science had become, once again, a battleground between left and right. This book is an attempt to reconstruct the complex relationship between political struggles and science in this epoch.

1
Between the Wars

In 1933 the Nazis came to power. Shortly afterwards, they began to reconstruct Germany according to their eugenic and racial ideals. This caused some concern among European and American intellectuals. A Professor I. Zollschan of Prague tried to mobilise his fellow intellectuals against Nazi race doctrine. He asked for and received the help of the Czechoslovak Prime Minister Masaryk to get under way an international scientific investigation of racial problems, with the object of countering Nazi propaganda. Zollschan was sent to various European capitals, with help from the Carnegie Institute and the support of the Czech President, Benes, to enlist the support of prominent scientists via the medium of the Institute of Intellectual Cooperation. In London in 1934 he contacted the anthropologist C.G. Seligmann at the London School of Economics. It was not the best choice from the point of view of promoting the brotherhood of man. Seligmann was firmly of the opinion that the black races were inferior to the white. But he was Jewish and he helped Zollschan as well as he could. He wrote, on his behalf, to Ruggles Gates, at Kings College, London:

> This is to introduce Dr I. Zollschan of Prague who has for many years, been interested in the question of race and character. He comes over with introductions from Masaryk, has succeeded in interesting the French and I understand has the Carnegie Institute behind him. He has just issued a short pamphlet which brings into doubt many of the absurd racial doctrines which are being taught in Germany at present. He desires to organise a discussion in which representatives of various countries and universities shall join and perhaps the matter may be brought up at the International Congress of Anthropological Sciences.[1]

Zollschan's mission was doomed to failure. According to an account by A. Metraux of UNESCO in 1950 it was sabotaged by 'the representative of one of the great powers' who 'frankly opposed the plan on the grounds that Hitlerite Germany must not be affronted'.[2] Zollschan was a believer in progress and reason. He believed that questions of race could be settled by 'a public appeal to science'. He hoped that a scientific enquiry would 'aim at outlining the degree of scientific certainty or uncertainty in the theories in question so far as that can be settled at present by a discussion between selected representative exponents of opposing views'.[3]

Unhappily Zollschan's confidence in reason and debate was misplaced. Many anthropologists were not believers in racial equality. In other instances, political considerations took priority over moral or even intellectual ones. German anthropologists or racial scientists did not experience isolation in the 1930s. None the less the direction the biological sciences were taking in Germany was a cause for concern in scientific circles. In 1936 the British scientific periodical *Nature* editorialised on a meeting by the British Association for the Advancement of Science at Blackpool, which had discussed the impact of genetics on traditional physical anthropology. This was the reason for 'a joint meeting . . . at which the speakers were primarily concerning themselves with the question of whether or not the anthropologists' definition of race required modification in consequence of recent developments in genetics. But no-one at this crowded meeting was unaware that the speakers were indirectly commenting upon the explanation of the race concept by politicians who apparently are deliberately confusing linguistic terms such as "Aryan", cultural terms such as "Germanic" and genetic terms like "Nordic" by using them synonymously'.[4]

Racial ideas such as these, claimed the editorial, had led to calls for territorial adjustment, immigration quotas to fit the racial divisions of Europe and to the belief that one national group had superiority over others. *Nature*'s editor was hostile to Nordicism. In Paris, the review *Races et Racisme* began publication in 1937 with the object of informing the public about Nazi racial doctrines and attacking them. But there was little concerted attempt to deal with these questions on an international basis. The real international impact came from the increasing distance between genetics in the West and the Soviet Union.

The Soviet view on genetics and heredity was largely determined by the idiosyncracies of her agricultural development. The rise to prominence in the 1930s of the Soviet agronomist Lysenko was largely because his 'alternative' genetics — in fact it was a revival of *Lamarckianism* or use-inheritance — was claimed to be able to increase rapidly the productivity of Soviet agriculture. But lacking detailed knowledge of these events, friends and foes of the Soviet Union alike tended to see the ructions which the Soviet government caused at international genetics conferences in the late thirties as purely a response to the rise of racial thinking in Europe. The foes of the Soviet system were often delighted that the apparently inegalitarian character of genetics offended Soviet doctrine and sometimes attempted deliberately to embarrass the representatives of the Soviet government at these genetics conferences by, for example, insisting on sessions on eugenics. But those more friendly to the Soviet Union believed one way to end its increasing self-imposed isolation in genetics was to restate the science in a strongly anti-racist form.

In 1932, the International Committee of the Genetics Congress decided to site the next one in Moscow in 1937. But by 1936 the tensions in Soviet genetics had reached boiling point, as had Soviet fear and suspicion of foreign involvement in their affairs, and the congress was cancelled. When criticism was broached about the curtailment of freedom of discussion of genetics in the USSR, the Soviet reply was prompt: 'There really does not exist in the USSR that "freedom" of genetical science which in certain states is understood as freedom to kill people or as freedom to destroy whole nations because of their alleged "inferiority".'[5] In addition, reports began to circulate in Europe and America in 1936 that the dismissal of S.G. Levit, director of the Maxim Gorky Medico Genetics Institute, had been, it was claimed, because he had said the Buryat peoples of Siberia had a mental age of twelve.[6]

Information about political conditions in the USSR was neither good nor soundly based. Indeed the Western scientists friendly to the country read many of these events through the medium of their own preoccupations about the rise of Hitlerism. It led them in 1939, at the rescheduled genetics congress at Edinburgh, to try and persuade, unsuccessfully, a Soviet delegation to attend, by having an anti-racist and anti-eugenic manifesto. The original signatories to this were F.A.E. Crew, J.B.S. Haldane, S.C. Harland, L.T. Hogben, J.S. Huxley, H.J. Muller, and J. Needham. The manifesto was a response to the question posed by the editor of *Science Service*: How could the world's population improve most effectively genetically? In the opinion of the framers of the manifesto no improvement was possible in conditions of racial antagonism.[7]

> The removal of race prejudice and of the unscientific doctrine that good or bad genes are the particular monopoly of particular peoples or of persons with features of a given kind will not be possible, however, before the conditions which make for war and economic exploitation have been eliminated. This requires some effective sort of federation of the whole world based on the common interest of all its people.[8]

The manifesto went on to specify that economic security — a job and a good standard of living — was also good eugenic policy. It stated the right of women to work and the importance of welfare benefits to allow them to do so whilst having and raising children. It also stated that the whole range of reproductive technology then available — birth control, sterilisation, abortion, genetic counselling — should be available but on a strictly voluntary basis.

The manifesto had no effect. In the middle of the congress, war broke out and the delegates departed rapidly for their own countries. It took a war before concentrated international action on questions of race was finally undertaken or before eugenic policies were re-examined in the light of events in the 1940s.

The advent of Hitler also disrupted other scientific communities. In the 1930s physicists continued to make significant discoveries in nuclear science.[9] Dotted across the world were centres of advanced research in atomic physics. These included the Cavendish Laboratory at Cambridge, centres in the USA at Columbia, Chicago and the University of California. In Copenhagen, Nils Bohr headed an important research team. In Berlin, Otto Hahn, Lise Meitner and Fritz Strassman worked at the Kaiser Wilhelm Institute. In Italy, Enrico Ferri and his team were making discoveries in this field. In the USSR, the Russian physicist, Kapitza, a former pupil and colleague of Rutherford at the Cavendish began, from 1934 when he was forced by Stalin to remain in the Soviet Union, to build up atomic physics in the USSR. In France, the Joliot-Curies headed a programme in experimental physics. There were other less famous but significant centres of research, sometimes comprising one or two individuals in the field. However, the circle of atomic physicists was small enough for each to be well aware of each other's work and incorporate each other's discoveries into their own experimental programmes relatively quickly. This awareness created a shared enthusiasm sometimes tinged with a competitive spirit. It was cemented by personal contact, through international conferences and because many had passed through the Cavendish or other great laboratories as pupils before moving on to set up or join other research institutions in atomic physics.

Two things threatened this network of scientists in the 1930s. In Germany, Jewish scientists had long experienced anti-semitism within the university system. But with the accession of Hitler to power in 1933, anti-semitism was encouraged and eventually was given the force of government decree. With this began the long trek of Jewish scientists across Europe, joined in many instances by scientists compromised in Hitler's Germany by their left-wing views. Leo Szilard, the Hungarian physicist, left Berlin in 1933 going first to Britain and then the USA. Otto Frisch and Lise Meitner left Germany, first for Bohr's team in Copenhagen. They were then forced, along with Bohr himself, to flee the German advance into Denmark. Frisch went to Britain, Meitner to the USA. Ferri in Italy was eventually forced to leave by events there. The closeness of the community of atomic scientists meant that, although there were cases of hardship, many of the fleeing atomic scientists found work in the same field. This led to a kind of cross pollination of research from which the host countries benefited greatly. But the other effect was to create political divisions inside the scientific community, especially when it became clear that atomic physics might eventually have a practical outcome in the form of energy for an atomic weapon. At this point not only did atomic physics take on a new meaning for the anti-Nazi but also for those among the atomic physicists who saw, in an approaching war, that their own nation's strength would be tested in battle. National as well as political considerations came to play a part in the calculations of the scientific community.

Prior to 1938, there had been tentative discussions about the possibility of harnessing atomic power for both economic and military purposes. But eminent physicists such as Rutherford were sceptical. In any case, these speculations were not as yet based on experimental results. The turning point came in late 1938 with the discovery of fission and it came in Berlin. Thereafter most scientists felt that, even if there was no immediate possibility of creating usable atomic power, there was a clear long-term possibility. Because of this, the outbreak of war in 1939 was a momentous event for many atomic scientists.

Atomic scientists were the first to alert their respective governments to the possibilities now opening in atomic science and also of the need to secure the raw materials — uranium and heavy water — which could be used in the manufacture of an atomic weapon. There were indications in 1939 that Germany itself was taking similar precautions. In 1940, two scientists, Otto Frisch and Rudolf Peierls, who were refugees from Hitler, sent a memorandum to the British government stating the possibility that an atomic bomb could be manufactured. In fact steps had been taken in 1939 by the British government, on scientific advice, to secure uranium and prevent heavy water manufactured in a plant in Norway from falling into German hands. The Frisch-Peierls memorandum however, led to the British government devising a programme for the actual construction of the bomb.

Similar steps were taken in America. By 1943 the British and US governments were co-operating in the production of an atomic bomb, the bulk of British research and scientists being moved to the Manhattan project at Los Alamos. Close communities of scientists were recreated but this time along strict national boundaries. The Joliot-Curies, still in occupied France and involved in the Resistance, dropped work which had relevance to the bomb and concentrated on other aspects of atomic research. Frederic Joliot-Curie's chief interest was in the economic benefits for industry of harnessing atomic power. Concealment and strict security became the order of the day everywhere but this change was generally acceptable, given the war. How far this would apply once it had been won was a different matter.

British scientists were affected by these developments but they also had more parochial concerns. The inter-war years in Britain were a low point for scientists. Whilst substantial and world-shaking discoveries were being made in pure science at the Cavendish Laboratory at Cambridge and elsewhere, the social position of scientists in Britain was insecure. Britain, relative to Germany and the USA, was an 'unscientific' community as measured by the criterion of the numbers of graduate scientists the country produced.[10] In part this was due to the structure of British industry, which would have found it difficult to employ graduate scientists even if the education system could have been encouraged to produce more. Britain had been slow in starting some technologically advanced industries such as chemicals and electrical engineering and even when these industries were established, Britain tended to specialise in heavy chemical and heavy electrical engineering. Compared to Germany, the process of monopolisation of industry was slow.

Whatever its other disadvantages monopolisation led to a greater concentration of resources, more comprehensive forward planning and greater investment in research and development. In addition, Britain traditionally recruited technicians from the shop-floor rather than from the universities.

The typical path for a scientist in industry was often apprenticeship to a skilled trade and shop-floor experience. This, as employers knew, meant a lot in terms of skill and experience but it also meant a lack of training in academic science and newer technological innovation. The skills acquired by the traditional route tended to be job-specific. The traditional route also affected the status of industrial scientists. Technical and scientific jobs were associated with the shop-floor rather than with management, with blue-collar origins. So in the 1930s, more graduate scientists were employed in education than in industry. When industry, after 1945, began to demand more graduate scientists, the education system responded but the position of the university-trained scientist in industry was still an uneasy one. Employers' views of their role were still structured by old-fashioned expectations. Thus there were frequent complaints that their knowledge was irrelevant to the job in hand and that they had no shop-floor experience but, none the less, higher expectations of status and reward than the technician trained by the companies themselves. These overall problems did not mean that either 'pure' science or craftmanship and technical skill were not in a healthy condition in the Britain of the 1930s. But it underlined an often expressed feeling of alienation experienced by academically trained scientists in industry. This situation had led to several things. One of them was the growth of scientific trades-unionism.

The National Union of Scientific Workers (NUSW) was founded in 1917 and registered as a trades union in 1919. The need for some form of professional body concerned with wages and conditions was felt during the First World War when increasing numbers of scientific workers were drafted into industrial and government service. The experience of scientists during the war convinced some of them of the need for a body to protect scientists' interests. Although the inspiration came initially from academic scientists, the thrust of future development was seen as embracing scientists in non-academic jobs and of every grade. In fact the NUSW was a response to the growth of a white-collar (or white coat) proletariat. The same response produced, at this time, other trades unions of white-collar and professional strata. These were distinguished from older trades unions by the educational qualifications of their members but they often experienced similar problems.

The NUSW was resisted to a degree by scientists because of its trades union character. Many scientists thought of themselves as 'professional' and shunned unionism. In recognition of this the NUSW became, in 1928, the Association of Scientific Workers (AScW). During the 1920s membership was small. It fluctuated between 1,000 to 2,000 members out of a potential membership estimated by some as 20,000. In 1934 under the impact of the depression, membership dropped to 695.

The AScW never saw its role solely in terms of pay and conditions — although its officials complained that its membership often did. It believed it had an additional purpose, to campaign to raise the status of science in society as a whole. In the 1930s it fell much more under the influence of the left and this left-wing influence survived into the post-war world.

The war brought better years for scientific trades-unionism. Along with the fall in unemployment went a general rise in trades-union membership from which the AScW benefited. Patrick Blackett, who was president from 1944–7, recorded a total of 19,000 members in 1947, of whom over half were fully qualified scientists and the remainder technicians and students. The AScW built up strength in industry in these years. Engineering and chemicals accounted for 11,133 out of its 18,949 membership; metallurgy for 1,130; the Civil Service (industrial and defence work) 1,215; university staff for 1,266.[11]

The status of trades-unionism improved generally in Britain in the 1940s due to the bargain struck by Ernest Bevin formerly of the Transport and General Workers' Union (TGWU), who during the war was Minister of Labour. Increased negotiating rights and recognition were exchanged for no-strike agreements and wage restraint. The growing importance of the Trades Union Congress, representing the trades union movement as a whole, led the AScW to affiliate with it in 1942. Thereafter much closer co-operation ensued between the AScW and trades-unionism in general — one significant outcome of the 1940s.

The post-war year saw the further development of scientific trades-unionism with the foundation in 1946 of the World Federation of Scientific Workers (WFSW) on the initiative of J.D. Bernal. The objects of the WFSW were to promote the interests of scientists world wide, much as individual associations did for their national membership. The WFSW was also very much of a popular front organisation transplanted to the 1940s, peace and international co-operation being high on its agenda. Much of its work, on tuberculosis for example, was uncontroversial but its position was compromised in the eyes of many by the domination of the Eastern bloc in its membership. Although the USSR did not join until 1951, the WFSW's largest membership was from Hungary, China (after 1949) and the German Democratic Republic (East Germany). Including the 20,000 members from the Soviet Union, these countries accounted for 79,000 of its 150,000 members in 1954. The French and British contingents, by affiliation of their respective national scientific organisations, accounted for a further 26,100. The motivating force in Britain was Bernal and, in France, Frederic Joliot-Curie, both very close to their respective Communist parties. The USA in contrast had only 250 members of the WFSW.[12]

But the ultimate solution for the problems confronting science was to find it a more secure role in national life. During the inter-war years intellectuals interested in science gave thought to this problem. In these years, discussion moved from a kind of optimistic utopianism in the 1920s, when all forms of possible scientific utopias were discussed, to the more serious programme of scientific reconstruction put forward in the 1930s by the Marxist disciples of J.D. Bernal. In the intervening years, however, it was Mondism, which gripped the imagination of many scientists interested in political and economic change.

Alfred Mond was head of Imperial Chemical Industries. This giant chemical firm had been founded by the Mond and Brunner families in Lancashire in the nineteenth century. It was exceptional in British industry on two counts. First it had grown large by the amalgamation of firms in the chemical industry. It was, in fact, as near a monopolised concern as could be found in this period. Second, it was technologically advanced, committed to research and development and it employed a great many scientists in its extensive research laboratories. The slump of 1929–31 affected it, but it survived better than many other industries. Alfred Mond, son of the original founder of ICI had strong views on industrial relations. In the 1920s, the phrase 'Mondism' described not only Alfred Mond's advocacy of large rationalised research-orientated businesses but also his view of the relation between the employer and the employee. Mond's views were, to a great extent, determined by the structure of his industry. It had few competitors, large resources and a good share of the market. Therefore in contrast to the frantic wage cutting which took place in the 1920s in the coal and textile sectors where firms proliferated, ICI could take a more lofty view of the relationship between employers and exployees largely because its survival was not so threatened. Mond believed in joint negotiating machinery between the two sides of industry and industrial peace based on union-management co-operation. After the General Strike of 1926, Mond, in alliance with some other captains of industry and prominent trades union leaders, inaugurated the Mond-Turner talks which tried to re-establish a degree of union-management agreement and co-operation in British industry.

Groups of thinkers associated with the Progressive League, the short-lived review *The Realist* (funded by Mond) and what passed for 'modern' thought in the 1920s found Mondism a congenial ideology. It suggested a way forward towards a stronger and more productive capitalism, one which utilised science and the scientifically trained worker. The Institute of Industrial Psychology founded in the early 1920s saw its role as complementary to the process of industrial rationalisation. It would introduce science into production and organisation of the large capitalist firm. It would train executive manpower in the principles of worker management.

The Eugenics Society too dallied with the idea that selective breeding could help in the construction of a new industrial order. In an article in 1934 on the progress of science, Julian Huxley cited industrial rationalisation and psychological and eugenic selection as the means of ensuring Britain's industrial progress.[13] The editor of the scientific periodical *Nature*, Sir Richard Gregory, was also a strong proponent of these ideas.[14]

The 1930s, however, brought changes. However rationalised German industry had been in comparison to Britain this had not insulated her from the economic slump of 1929–31.

The depression of the 1930s suggested that rationalisation alone could not save industrial capitalism. In fact some argued that the high unemployment of the 1930s *was* rationalisation in practice. If this was the case, rationalisation did not augur the prosperous and expanding economy which Mondism had promised. On the contrary it seemed to co-exist with high unemployment, poverty and stagnation in large sections of British industry.

Similarly Mondism was inimical to militant trades-unionism. The miners' leader Arthur Cook unflatteringly referred to 'Mond's manacles' in his pamphlet of the same name. Moreover, the principle of psychological and eugenic selection as the means to fit people into industrial and social slots came under increasing criticism in the 1930s. The overt race and class prejudice in eugenic circles caused some scientists to back off the idea of using it as a means to construct a new social order.

The economic depression of the 1930s influenced some towards a more favourable view of the Soviet Union which, however backward it seemed in many areas, could point to other areas of very rapid development. The idea of the large industrial monopoly which could plan ahead and use the resources of science still held sway but under the influence of J.D. Bernal it acquired a socialist reinterpretation.[15] Bernal was born in 1901 at Nenagh in Ireland to a Catholic family. Most of his education was in England, and in the 1920s he worked as a crystallographer at Cambridge. In the 1930s he moved to Birkbeck College, London.[16]

Bernal had gone through a 'utopian' phase publishing in 1929 a Wellsian fantasy, *The world, the flesh and the devil* which earned him brief notoriety. He had been closely associated with the Communist Party since about 1923. It was not, however, until the early 1930s, in particular the visit of Bukharin and Hessen as representatives of the Soviet government at a conference on the history of science in London, that his ideas moved from cloudy utopianism to a more rigorous theoretical Marxism.

In the 1930s Bernal reinterpreted Mondism to his scientist colleagues. Bernal accepted the argument that, historically, capitalism had unleashed tremendous productive forces and this included great advances in science and technology. In pursuit of this idea Bernal produced lengthy histories of scientific development. In these, taking his cue from the Soviet Marxist Hessen's description of the origins of Newton's *Principia* as part of the seventeenth-century expansion of commerce, he linked science with the development of capitalism. Bernal also adapted the Mondist view that amalgamation and large-scale planning was necessary for future progress. However, he believed capitalism could no longer fulfil this role. Crippled by the inner irrationality of competition and financial crisis, capitalism had surrendered to socialism its role as midwife of scientific change. Socialism could provide the rational economic planning which the crisis of 1929–31 had checked.

Not all of these ideas could possibly attract or appeal to scientists as a whole. But during the 1930s the influence of these ideas increased, not just among left-wing scientists. This was partly due to the impasse in which scientists found themselves in the 1930s with dwindling professional opportunities and what seemed a largely indifferent government. So even if they were not prepared to go as far as the full-scale state planning advocated by Bernal, many were prepared to demand some form of re-organisation and planning in industrial and government research.

Bernal's ideas had sufficient influence to seriously alarm some scientists. In 1940, a chemist Michael Polanyi and an Oxford zoologist John R. Baker circulated a letter to 40 scientists who they felt would be sympathetic to some form of association to combat Bernalism. In particular Baker had been disturbed by speeches and debates at the British Association for the Advancement of Science in 1939 and at the Royal Society which he felt were too Bernalist in tone. Out of this venture came the Society for Freedom of Science (SFS) whose aim was to combat the idea that the future of science lay in large-scale planning and government direction. The year, 1940, was not propitious for its advent but in the post-war world the SFS was to play an important part.

Meanwhile other more urgent international issues began to press in on British scientists. Not many of them were pushed towards the left in the 1930s but those who were played a vocal and influential part in the subsequent history of scientific politics. J.B.S. Haldane, the biologist, had been a founder member of the Cambridge University Liberal Club shortly after the war and a member of the Cambridge Eugenics Society. In the late 1920s he was 'progressive' minded but still hovering about the liberal centre. By the outbreak of the Spanish Civil War in 1936, Haldane had become a socialist and had moved close to the Communist Party.[17]

In the 1930s, he was joined in various campaigns by other scientists, some of whom had long-established links with the left. Hyman Levy, for example, had had, like Bernal, associations with the Communist Party since the early 1920s. The biologist Lancelot Hogben's political beliefs were much more erratic and unpredictable but he was certainly of the left. Patrick Blackett, a physicist, who worked at Rutherford's Cavendish Laboratory in the 1920s was a member of the Labour Party. But whatever the particular political orientation of these and other scientists in the 1930s they worked together in various institutions and campaigns to the extent that they can rightly be called, as Gary Werskey has done, a 'visible college' — an inter-connected network of the left.

The campaigns in which they participated were the usual ones which attracted the left in the 1930s but, in addition, there were special areas of concern to scientists or ones on which they felt scientists should speak. One group used for this was the AScW which, after the membership slump of the early 1930s, was slowly beginning to recover. Through the medium of the AScW and in the British Association for the Advancement of Science (BAAS) and elsewhere, they put the case for science. Hogben and Haldane, in addition, worked in the 1930s against the racial and class bias they felt existed in the eugenics movement in Britain. Haldane particularly, through the medium of the socialist press, helped keep genetics in tune with a healthy popular egalitarianism. In addition most of these scientists were concerned with 'popular front' politics, that is resistance to the advance of Fascism in Europe. Through the studies of air raids and civilian bombing undertaken by Haldane in Spain and carried on by the Cambridge Scientists Anti-War Group in their backyards, they indirectly contributed to the planning of civil defence on the outbreak of war in 1939. The war years saw major changes in Britain. The rapid fall in unemployment after 1940 and the total mobilisation of the population in the war effort affected scientists too. Talents which had been unused or underused in the 1930s began to be put to work. Although the security services were wary of them, the left-wing scientists of the 1930s were co-opted into the service of the government. Blackett worked under Mountbatten at the admiralty, utilising his First World War experience in the Navy. Bernal was taken on as an adviser to Sir John Anderson on civil defence and Haldane worked on military research.

These years were very congenial for Bernalism. The war-time government was committed to extensive intervention in the economy, including direction of labour and control of raw materials, and also to large-scale planning. Unemployment fell dramatically and production rose. In 1945, with the election of a Labour government, many of these controls were retained and the government was pledged to using them to plan post-war economic reconstruction. The AScW welcomed these developments. Blackett, who was president of the AScW, considered that the majority of its membership was sympathetic to the aims of Labour.

> But the fact is that in the first decades of the twentieth century the majority of the members (particularly the senior ones) of the older scientific organisations were in fact as soundly conservative as the rest of the professional middle classes. Now the younger scientists both in the universities, in industry and in government employ were increasingly coming to the view that their interest as scientists as well as the interest of the community as a whole lay with the programme of the socialists

or at any rate of the progressive political parties. It was roughly speaking scientists with 'left' political views who joined the AScW and it would be certainly true to say that now with the Labour government in power, the majority of members of the AScW would consider themselves as supporters of their general policy, while equally clearly the majority of the senior scientists in the older professional bodies support individually the policy of the opposition.[18]

Whether Blackett represented the political views of AScW members fairly or not it is none the less true that he had highlighted a fundamental difference within the scientific community. Not all thought that trades-unionism was the proper way forward even when they might agree that the status of science was low or the educational system insufficiently science-orientated. Trades-unionism pointed towards mass action and democratic decision making. It therefore cut across the idea that the best way forward for science was the permeation of 'elite' values with science. One proponent of this latter view was Lord Cherwell, formerly R. Lindemann, who was the adviser and friend of Winston Churchill both before and during the Second World War. Cherwell acquired significant influence over both scientific and economic affairs in this period. He had done this by joining the country house circuit and acquiring a life-style and attitudes reflecting it. He had made friends in the right places. In his case this had been highly effective, much to the annoyance of those scientists who had climbed the political ladder slowly by discreet and conscientious government service rather than by more flamboyant socialising.

Cherwell was anxious that science should avoid association with the left or trades-unionism. When he pondered the inadequacy of science education in the 1950s he anxiously discounted the idea of special technical schools. Cherwell was keen to see science accepted into the traditional hierarchical structure of British education — as part of the public school and Oxbridge system. Science should become part of 'general' education. Proposals for separate science-based schools and colleges would simply perpetuate the divisions between the scientists and the social and political elite.[19]

At the end of the war in 1945, therefore, all the elements were in place but another was added. By 1946 the tension between USSR, America and Britain was adding a new factor to the political concerns of scientists, one which was to affect them very profoundly.

Notes

1. Ruggles Gates Papers, 1930-5, Seligmann to Ruggles Gates, 15 March 1934.
2. A. Metraux, 'UNESCO and the racial problem', *International Social Science Bulletin*, vol. 2, no. 3 (Autumn 1950), p. 386.
3. I. Zollschan, '*The significance of the racial factor as a basis in cultural development*', (Le Play House, London, 1934), p. 2 and p. 16.
4. *Nature*, vol. 138 (12 December 1936), pp. 988-9. The meeting was at Blackpool, 1936.
5. The Genetics Congress, *Journal of Heredity*, vol. 28, no. 1 (January 1937), p. 55.
6. *Nature* vol. 139 (January 1937), p. 142.
7. 'Men and mice at Edinburgh', *Journal of Heredity*, vol. 30, no. 9 (September 1938), p. 371.
8. Ibid., p. 372.
9. See John Hendry (ed.) *Cambridge physics in the thirties* (Adam Hilger, Bristol, 1984); Lawrence Badash, Elizabeth Hodes and Adolph Tidden, 'Nuclear fission: reaction to the discovery in 1939'. Proceedings of the *American Philosophical Society*', vol. 130, no. 2 (June 1986) Alwyn Mackay *The making of the atomic age*, (Oxford, Oxford University Press, 1984).
10. R.M. Pike 'The growth of scientific institutions and the employment of natural science graduates in Britain 1900-60', MSc, University of London, 1961.
11. Blackett Papers E23, Association of Scientific Workers Papers, MSS79/ASW/1/2/29/1-3 (1948-9).
12. WFSW Papers MSS270 (Peter Wooster Papers).
13. Julian Huxley 'The applied science of the next hundred years', *Life and letters*, vol. 2 (1934), pp. 38-46.
14. Gary Werskey, *The visible college* (Allen Lane, London, 1978).
15. J.D. Bernal, *The social function of science* (Routledge and Kegan Paul, London, 1939).
16. M. Goldsmith, *Sage: a life of J.D. Bernal* (London, Hutchinson, 1980); A.H. Teich, *Scientists and public affairs* (Harvard University Press, Cambridge, Mass, 1973); J.D. Bernal 1901-71, in *Biographical memoirs of the Royal Society*, 26 London 1980; M. Goldsmith and A. Mackay (eds) *The science of science* (Souvenir Press, New York, 1964).
17. Ronald Clark, *J.B.S.: the life and work of J.B.S. Haldane*, (Hodder and Stoughton London, 1968); K. Dronamraju (ed.) *Haldane and modern biology* (John Hopkins, Baltimore, 1968).
18. Blackett Papers E23, 'The development of the Association of Scientific Workers' (1947).
19. Cherwell Papers J146, 'Memo on the availability of trained scientists for rocket research' (1955).

2
Proletarian and Bourgeois Science

In August 1948, Soviet biology became officially Lysenkoist. On 11th August 1948, Lysenko addressed the Soviet Academy of Agricultural Sciences. This was followed by a meeting of the Council of the Academy of Sciences of the Soviet Union on 26th August. At these two meetings the direction of Soviet biology was substantially transformed. A number of prominent geneticists lost their posts or were transferred into other work. Lysenko, who since 1939 had been president of the Lenin Academy of Agricultural Sciences, became director of the Soviet Academy's Institute of Genetics. The decrees adopted at these meetings ordered Soviet research institutes and institutions of higher education to adopt the principles of Michurinism. Michurinism developed from the thinking of the Soviet plant breeder Michurin whose disciple Lysenko considered himself. According to Lysenko, Michurin's work developed the practice of environmental determination of heredity.

Recent work on the adoption of Lysenkoism in the Soviet Union attributes its victory to Lysenko's ability to convince Stalin and other Soviet officials that his methods could lead to a rapid improvement in the productivity of Soviet agriculture. Its fall from favour in the late 1950s was the result of its failure to fulfil these promises. At the same time Lysenko's allies such as the philosopher, Prezent, attacked Mendelism as idealist and undialectical.[1] This was not the decisive factor in Lysenkoism's triumph but scientists outside the Soviet Union, whose knowledge of Soviet agriculture and its difficulties was limited, were inclined to see the philosophical battle between Mendelism and Lysenkoism as the significant one.

In Britain, the Lysenko controversy had three effects. It helped break up the alliance of left-wing and radical scientists which had emerged in the 1930s. It was one of the factors which allowed the

British government to exercise a stricter and more rigorous ideological and political control over the scientific community at a very significant moment in the government's relationship with science. Finally it affected the cultural milieu around science. It allowed re-definition of the role and character of science.

The development of atomic weapons made the question of scientists' political attitudes and their co-operation with government of great importance to the governments of the USA and Britain in the years after the Second World War. An episode like the Lysenko controversy which, before the official endorsement of Lysenko by the Soviet Union, had seemed bizarre and esoteric to most British scientists was to assume considerable importance in the very different atmosphere of 1948. In 1947 Ernest Bevin, the British Foreign Secretary, had placed a memorandum before the Cabinet setting out the deterioration in the international situation.[2] It called for an effort to publicise the British way of life and to attack violations of human rights in Eastern Europe. The Lysenko controversy became one of these violations — part of the cultural cold war. But it also drove a wedge between the scientists themselves. It became part of the means by which the radicalism of a section of the scientific community was recast into the conservativism of the 50s.

There was no pressure group for Lysenkoism among British scientists before 1948. Eric Ashby's observations on Soviet science before 1948 were typical of the general attitude to Lysenko and Soviet biology. In a favourable review of Hudson and Richen's book on Lysenko in 1947, Ashby described his work as 'militant obscurantism', his experiments as 'lazily and carelessly constructed' and his terminology as 'quaint and picturesque'. On the whole, however, Ashby concluded that:

> As to the present state of the new genetics in the Soviet Union, it is safe to assume that Lysenko's school is well past its zenith. It is true that he is still a great demagogue among the peasants on the collective farms. It is true that, in the University of Leningrad, Turbin (who has the chair of genetics) and Prezent (who has the chair of Darwinism) maintains a school of the new genetics . . . But side by side with the new genetics, there exists in uneasy truce, a school of the old genetics which is setting the pace in world standards in such fields as population genetics.

Ashby went on to argue that, 'It is fairly safe to say that most biologists in the USSR . . . are deeply embarrassed by the new genetics and are gradually turning their back on it.'[3]

This opinion was shared by K. Sax writing in *Scientific Monthly* in 1947. He believed that 'though most farm workers, agronomists and political leaders, subscribe to Lysenko's doctrines some good work is being done in genetics in the Soviet Union. The general philosophical tribute to Marxism required by biologists was not necessarily proof of acceptance of Lysenkoism.'[4] Dobzhansky too considered that, 'We may be confident that the brilliant and active group of geneticists now working will keep the USSR in the forefront of scientific progress.'[5] Julian Huxley maintained that in spite of Lysenko, 'The approach in genetics has, in general, been along the same neo-Darwinian lines, involving acceptance of neo-Mendelism as in Britain and the United States though perhaps with even greater emphasis on strict selectionist principles.'[6] Huxley also pointed out the existence of Mendelians within Lysenko's own Institute.

Were these scientists deceiving themselves because of the generally mild attitude to the politics and culture of the Soviet Union produced by the war alliance? Though this may have played a part, it could not be held as primarily responsible. For example, even J.R. Baker and A.G. Tansley of the Society for Freedom in Science believed in 1946 that the autonomy of science was increasing in the USSR. It was true they wrote that, 'In certain fields of research such as genetics, Soviet standards and criteria are almost incredibly perverse',[7] but after quoting from Hudson and Ritchens they continued that, 'It may be that the inevitably disastrous effect on practical results in the long run of such a travesty of sound scientific method is the cause of the change of policy apparent in the articles of Kapitza and Joffe.'[8] Judging by Joravsky's recent book on Lysenko, these opinions were, by and large, correct. Geneticists had continued their work. Lysenko's influence was by no means paramount and he had not, at that stage, secured the position of influence with the Soviet government he enjoyed after 1948. But certainly analyses of Lysenko prior to 1948 conveyed an overoptimistic view of the future of biology in the Soviet Union.

There were two exceptions to this. The Society for Freedom in Science were critical of Lysenko's influence even before his victory in 1948. They regarded his emergence as evidence of the distortions socialism brought about in science. From a different political perspective, the prominent Communist biologist J.B.S. Haldane made unfavourable comments on Lysenko's work in the 1930s.

Although he framed his criticism in a friendly and fraternal spirit, Haldane disputed the claims made by Lysenko for his work.

Haldane's connection with Lysenko went back a considerable time. For example, Haldane found himself under attack from Polyakov — a colleague of Lysenko — for his attendance at the Conference on Genetics and Selection in 1939. Polyakov attacked a number of foreign scientists including Fisher, Wright and Haldane:

> I have referred to Haldane as a Darwinian geneticist and I believe I shall continue to do so. But we must not only study but correct them. When these Darwinian geneticists employed a statistics far more interesting than those we have been discussing, when they begin to build a whole evolutionary conception on this empty purely abiological statistics, then we can show that in the works of Haldane and the others there is no consideration of real biological interrelations of all that is concretely involved in evolution and much is transformed into an abstract scheme [sic].[9]

These proceedings were republished in *Science and Society* and Haldane replied to the criticisms of him in the next issue. His reply, whilst polite, was a vigorous rebuttal of Lysenko. It included most of the subsequent criticisms which geneticists were to make of Lysenko: his controls were not strict enough to justify the conclusions he drew from his experimental work; he had charged geneticists with beliefs — such as the immutability of the gene — which they did not in fact hold; most important of all, he had misunderstood the role statistics played in the theory of evolution attributing to it a kind of metaphysical role it did not in fact possess. However, though Lysenko had gone too far, 'He may well have done a service to Soviet genetics by making his more traditionally minded colleagues examine not only the theoretical foundations of their work but its relation to agricultural practice.'[10]

In 1947, signs of an upheaval in Soviet biology were observed and discussed in the pages of the socialist periodical *Modern Quarterly* by R.G. Davies, J.L. Fyfe and John Lewis. Lysenkoism on the whole received an unsympathetic reception. Both Davies and Fyfe came down against him although later Fyfe changed his position publishing a defence of Lysenko, *Lysenko is right*, with Lawrence and Wishart (1950). In 1947, however, he regarded Lysenkoism with scepticism: 'On the scientific side of the controversy then, we should withhold judgement on Lysenko's experiments until they have been adequately tested, while the theories based on them may be disregarded as a contribution to biology.'[11] The Communist Party theoretician John Lewis conceded that opinion among biologists — including socialists — was against Lysenko. He asked only that Lysenkoism should be seriously examined before being dismissed and that the worse accusations against him — including responsibility for the death of Vavilov — be discounted.[12] Vavilov was one of the most important Soviet geneticists. He died in 1943, in prison.

The dominant tone of these articles was genuine puzzlement about the significance of Lysenkoism. The aspects of Soviet development which had produced Lysenkoism were not clearly appreciated outside the USSR. As yet there was no authoritative voice from the government of the USSR in its favour. Haldane forged the intellectual links between Marxism and biology and he was anti-Lysenkoist. As Haldane after 1948 often pointed out, the natural theoretical supporters of Lysenko were the Lamarckians, the Vitalists and the anti-Mendelians who were left stranded in various academic posts by the advance of genetics. They were often politically quite orthodox.

The socialist George Bernard Shaw was not politically orthodox but he supported Lysenko because he saw him as a fellow vitalist. Shaw regarded the 'dialectics' in Lysenko as an irritating and unnecessary residue of Marxism. Lysenko, he argued, 'Is on the right side as a Vitalist but the situation is confused by the purely verbal snag that Marx called his philosophy "Dialectical Materialist". Consequently Lysenko has to pretend that he is a Materialist when he is in fact a Vitalist.'[13]

In his article in *Modern Quarterly* in 1948 (published shortly before the impact of Lysenkoism began to be felt), Haldane set out his views on the philosophical relationship between Marxism and biology.[14] The interest of this lies in the fact that it was on this ground — the relation between dialectical materialism and biology — that the supporters of new 'genetics' frequently defended their work. The question of this relationship became one of the main areas of debate among European and British Marxist intellectuals.

Haldane found quite different links between biology and Marxism to those suggested by Lysenkoists. In 1948 he rested his contention of a connection between Marxism and biology on two main points — the materiality of living systems and the dialectical character of biological processes. Haldane admitted that biology was expressed sometimes in mechanistic and sometimes in idealist language. However biologists were propelled by the development of their discipline towards a recognition of the material character of life. The dialectic Haldane found in the antagonistic character of living processes. For example, he pointed out two opposing tendencies in living systems; the tendency to vary, without which evolution would be impossible, and the tendency to remain the same, which made heredity possible and which was, again, a condition of evolution. These principles were, in his opinion, perfectly compatible with Mendelism and Haldane did not think it worth while even to raise the possibility of conflict between them.

Ironically, the intellectual links between British Marxism and Lysenko were forged by the right rather than the left. It was for example the Society for Freedom in Science which suggested interconnections between Marxist philosophy of science and the Michurinist trend in the USSR. Secondly, a number of biologists appeared to give credibility to Lysenko's contention about a link between racism and genetics by their remarks on the origin of racial and social inequality. When Lysenko accused Mendelian genetics of being racist and supporting theories of human inequality, these biologists concurred. Mendelism, they argued, did make vain hope of a more egalitarian world.

The Society for Freedom in Science was founded in 1940 by John R. Baker, then a lecturer in zoology at Oxford. In an introductory letter, Baker outlined the reasons why he felt the Society was necessary. Primarily the SFS was aimed at refuting the idea that scientific progress was and ought to be determined by society's material wants. This idea, according to Baker, would produce a situation in which, 'Unless active steps are taken, there is a danger that creative research workers may be swamped under a deluge of mass-produced scientists who still clamour for work in organised groups on dictated problems immediately applicable to human affairs.'[15]

The Society for Freedom in Science, although it claimed in its constitution to be non-political, felt its chief task was to counteract the influence of radical scientific pressure groups. It saw its role as the focus of opposition to the 'widespread and vigorous movement which sees the solution of social difficulties in a complete recasting of the structure of society under a system of central control'.[16] Later on, in 1953 when Baker recounted the origins of the SFS and republished the letter of 1940, he was more forthcoming about the institutions against which the SFS intended to produce counter propaganda; in the potted history he produced in 1953, he cited the division for Social and International Relations of Science founded by the British Association in 1938, the Association of Scientific Workers, the meeting of the Royal Institution of 1941 and the work of science journalists in liberal newspapers.[17]

He was also more specific about the philosophical bent of the Society. 'From 1931 till 1938', he complained, 'Marxist writers had been able to publish whatever they liked about science without danger of their remarks being challenged.'[18] He mentioned the refusal of *Nature* to publish the aims of the Society as evidence of this left orthodoxy (although ironically he got the *New Statesman* to publish his manifesto on the 29th July 1939).

The protagonists of the SFS found themselves in an ambiguous position during the war. Its chief philosopher M. Polanyi, for example, touched on Lysenkoism in an article written in the 1941–3 period and published in 1945. Lysenkoism was cited as one of the distortions arising out of the influence of Marxist philosophy and the direction of research by the State. But these criticisms tended to be muted in comparison with those of the post-war years. The Soviet Union was, after all, a war ally. Polanyi stressed that the Soviet Union was by no means as bad as the Nazi regime. 'The attempts of the Soviet government to start a new kind of science are on an altogether different level. They represent a genuine effort to run science for the public good and they provide, therefore, a proper test of the principles involved in such an attempt.'[19]

Polanyi put forward what was, on the whole, the best critique of Lysenkoism from the right. If state support is a criterion of scientific research, he argued, then the dialogue of science is inevitably transmuted into one of political authority. In addition if genetics was to be judged by the measure of practical utility then its condemnation was inevitable. In other words, Polanyi argued that there was an inherent logic in the relationship between a 'socially' responsible science and the emergence of Lysenko.

However, Polanyi's relatively calm and judicious examination of Lysenko was unusual. In contrast many of the myths about Lysenkoism in the USSR arose from the SFS. One of these was the idea that a generation of left or liberal scientists had been duped into minimising Lysenko's influence. Baker, for example, claimed that the facts about Lysenko had been made known primarily through the SFS. In fact those who brought information to Britain about Lysenko before 1948 were figures such as Ashby, Huxley and Haldane himself. In his reminiscences of 1953, Baker also claimed that in the *Nature* report of the Genetics Conference in Moscow in 1941, 'The degraded nature of the proceedings had been glossed over'.[20] This was a strange conclusion since in that report the presence of Lysenko was the main theme and his work referred to in unflattering terms. Baker also claimed that it was only on the initiative of the SFS that the 'lenient' attitude of British scientists to Lysenko was counteracted.[21]

There was more important indirect influence exercised by the right. When the Lysenko controversy broke, a typical justification put forward by many left-wing scientists for switching to Lysenko's support was the racist character of much of contemporary genetics. The views of a number of geneticists on the question of race and class helped feed a counter-reaction on the left particularly since Lysenkoism was often depicted as an attempt by the USSR to escape the racist implications of research in heredity. C.D. Darlington, for example, in 1947, when director of the John Innes Horticultural Institute, drew a picture of irreconcilability between genetics and socialism based on the presumption that, 'A government which relies on the absence of in-born class and race differences in man as the basis of its political theory was naturally unhappy about a science of genetics which relies on the presence of such differences.'[22] In a footnote to the article in which this appeared he quoted with disapproval Ashby's comments on the Soviet school curriculum (pre-Lysenkoist at that time). This included the principle of 'the inadmissiblity of extending the theory of natural selection to human society. Such an extension is classified under heresies as "Social Darwinism".'[23]

Darlington's views on the survival of Soviet genetics in 1947 flatly contradicted those of many of his colleagues. 'No foreign book on genetics or cytology has been translated since Stalin took control. Their notions of Western genetics and cytology are therefore always 25 years out of date.'[24] Moreover Darlington claimed — this was to become an orthodoxy after 1948 — that there had been wholesale liquidation of geneticists in the 1930s.

A degree of rewriting of the history of Lysenkoism occurred before 1948. It followed these lines: exaggeration of the importance of Lysenko's influence in biology before 1948; the assumption that the disappearance and death of geneticists (in particular Vavilov) was due to his influence; that the adoption of Lysenkoism by the Soviet government was due to an inherent ideological contradiction between Marxist theory and genetics which could be traced back as far as Marx and Engels themselves.[25] This view, which was that of only a relatively small conservative section of the British scientific community before 1948, became orthodoxy after it. As Joravsky's book shows, it became the conventional 'history' of Lysenkoism.

Information about the decision of the Soviet Academy of Sciences to support Lysenkoism reached the British press in late August 1948 but was mentioned only briefly. Preparations for the annual meeting of the British Association were under way but the case of Lysenko was not raised there. By November, however, it was given more coverage. In that month, the Royal Society severed its relations with the Soviet Academy of Sciences. Ashby appeared on a BBC Third Programme broadcast to talk about Lysenko[26] and the case began to gain increased press coverage. By December, after a second broadcast involving four biologists, S.C. Harland, R.A. Fisher, C.D. Darlington and J.B.S. Haldane,[27] the Lysenko controversy was receiving major coverage in the press and periodicals.

The political climate in which the question of Lysenko became known was of considerable significance. It was at the height of the Cold War: 1948 saw the Berlin Blockade, the Polish and Hungarian elections, the confirmation of a Communist place in their governments and a Soviet military presence in those countries. The civil war in Greece was being fought; the Chinese Communists were driving towards Shanghai; there were emergencies in Malaya and Vietnam. Within Europe the French and Italian Communists were fighting to maintain the gains they had made after 1945. The same year saw the beginning of an official TUC drive in Britain against Communists in the Trades Union Movement. On 15 March 1948, a purge of the Civil Service had taken place which involved the removal of a number of scientists and technicians from their posts on the grounds of their political allegiance.[28] Of the 30 purged from the Civil Service 17 were scientists. These included one from Harwell Atomic Research Station, six from the Telecommunications Research Station at Malvern, four from the Royal Aircraft Establishment at Farnborough, four from the Woolwich Arsenal, one from the Post Office and one unknown.

These events had a considerable impact on the relationship between government and scientists. The movement of left-wing and radical scientists which emerged in the 1930s had continued to develop in the early 1940s when even pro-Sovietism (with some exceptions) became acceptable. Certainly it was an era which witnessed a move to the left of which they could take full advantage. After 1946 the tide was moving in the opposite direction. In addition government intervention in science had become a major issue because of the atomic weapons programme the British government had undertaken. The state support of science for which the 1930s radicals had called was taking place but in a different direction from the way they had hoped.

On the other side of the political divide at the very moment when the USSR was attempting — through the Wroclaw Conference in Poland — to cement an alliance of intellectuals friendly to it, the Soviet Academy of Sciences had introduced a divisive issue — Lysenkoism — into intellectual debate. This reflected the increasingly tight control the USSR exercised over its intellectuals in the post-war world. The *Manchester Guardian* was quick to point out the effects of this in an editorial of 2nd September 1948, which also reflected the mounting anti-Communist sentiment in this period.

> The Wroclaw Conference coincided with a meeting in Moscow of the Presidium of the USSR Academy of Sciences (widely reported in the Soviet Press) which supplies a significant commentary on the hypocrisy of the Communist pretence to uphold intellectual freedom . . . Perhaps these British 'intellectuals' who at Wroclaw so enthusiastically swallowed the Communist bait will now direct the 'struggle' for intellectual freedom eastwards. Or would that mean fewer paid holiday trips?[29]

The increasing anti-Communist political tone of even a Liberal newspaper is illustrated by the *Manchester Guardian*'s comments on the emergency in Malaya. The paper described this as 'not a genuine Nationalist rising but an alliance between criminals and Communists (if in this case the terms can be held to be distinct)'.[30]

The Lysenko controversy rapidly became part of the cultural cold war but the scientific issues took second place to the political. Lysenko was yet another touchstone by which two political systems were to be judged and, also, two different philosophies of science. This put under pressure not only those scientists like Bernal and Haldane who were openly associated with the USSR but also those who were not but who had shared with them a number of ideas about the role of science and a willingness to cooperate with them in the pursuit of certain objectives.

In November came Ashby's broadcast. Ashby did not retreat from the opinions he had expressed earlier. 'The opinion of many observers is that this is nothing new. There has been, they say, a genetics controversy in Russia for a dozen years; heretical biologists have been liquidated before; bourgeois biology has always been suppressed in Soviet Russia. This common opinion is contrary to the facts and quite unjustified.'[31] Lysenko had been adopted by the Soviet government, argued Ashby, because Lysenko's 'practical ideals do in fact work'. Also he was a bridge between the peasant and the government. It was the problems posed by collectivisation and agricultural production which had caused his rise to power. It was not the outcome of a necessary link between Marxist ideology and Lysenkoism. On the contrary, 'It requires a good deal of sophistry to demonstrate that the new genetics is any more Marxian than the old.'[32] Ashby's broadcast was followed by one considerably more polemical in tone. Three of the four scientists who participated — Harland, Darlington and Fisher — as well as condemning Lysenko's biology assumed it as axiomatic that Vavilov had lost his life because of his views on genetics. R.A. Fisher, for example, repeated the charge that 'It had become known that under the impulsion of his influence many Russian geneticists, and these among the most distinguished had been put to death either with or without pre-treatment in concentration camps.'[33] Haldane's broadcast stood out from the others partly because its tone was unemotional. He denied that Lysenko had been responsible for Vavilov's death. Secondly whilst rejecting Lysenko's biology, he insisted upon treating it as though it merited serious discussion.

On 22 November, Sir Henry Dale — a member of the SFS — resigned from his position as corresponding member of the Soviet Academy of Sciences and produced a letter in explanation which was, though a personal statement, a classic exposition of the SFS's position on Lysenko.[34] As well as repeating the accusation that Vavilov had gone to his death because of his views on genetics, it compared the fate of geneticists in Russia to that of Galileo in the seventeenth century. Sir Henry Dale's letter of resignation received wide publicity. Shortly afterwards the Royal Society also broke off contact with the Soviet Academy of Sciences. Sir Robert Robinson, president of the Royal Society, referred to these events at the annual dinner of the Society in December 1948 at which Herbert Morrison was an invited speaker. By then the question of Lysenko's scientific merits or demerits had given way to the question of politics. The president, Robinson, drew a parallel between freedom of science and British democracy: 'We enjoy a heritage of untrammelled scientific research which has been won for us in past generations and is as dear to us as the political liberties secured by Magna Carta.'³⁵ However, he expressed an element of unease about the position of scientists in Britain:

> A real danger lay in the exaltation of 'utility' as a sufficient end in itself. Whether it took the form as reported from Soviet Russia or any other form detrimental to the pursuit of knowledge for its own sake . . . Actually no direct attack is likely here and should the unexpected happen it will certainly not be along the lines of compelling us to espouse some particular scientific theory or doctrine. Conceivably it could take the more subtle form of control of the character of our scientific work.³⁶

Herbert Morrison, who as Lord President of the Council was in charge of security, including the atomic weapons programme, also spoke. He had read Sir Henry Dale's letter and thoroughly agreed with its terms and spirit. 'It was a declaration which would live in history as a declaration of the rights of man (cheers).'³⁷ Morrison commented on the Society's decision to cut its links with its Soviet counterpart approvingly since 'the struggle in which we were engaged today is for the whole future of Western civilisation, was for the defence of a few simple fundamental values without which free men could not live.'³⁸ He added slightly ominously, 'In other countries we hear of scientists being proscribed and persecuted on account of alleged deviation from political dogma. Any British or other scientist who supports this sort of thing will soon cease to be a scientist or at any rate a scientist on whom reliance can be placed.'³⁹ However, Morrison did not demand a total depoliticisation of science. He appealed for the emergence of a new breed of 'scientist-statesmen' 'leaders of public affairs'.

Haldane in spite of his record as Lysenko's critic felt the impact of these controversies particularly strongly. The *News Chronicle* kept up a persistent onslaught on him. On 1st October A.J. Cummings, their political correspondent, referred to an article in the American magazine *Time* according to which, 'Professor J.B.S. Haldane, the English biologist who is a Communist, has been set a poser by the spurious theory of genetics recently proclaimed in Moscow by Comrade Lysenko . . . But the Professor is chairman of the editorial board of the *Daily Worker* . . . What will that enlightened organ of opinion do about Haldane if he challenges Lysenko and openly deviates from Moscow's politico-scientific line?'⁴⁰

Haldane in fact took up the challenge and replied in the pages of the *Daily Worker* on 1st November 1948, with a cautious refutation of Lysenko. There were further attacks by Cummings and a long article by Ritchie Calder on the 29th, that singled out Haldane on the eve of a second BBC broadcast which went out on 30th November.[41] On 1st December, the day after the broadcast, another item by Cummings accused Haldane of dodging the issue on Lysenko. Finally the *News Chronicle* gave space to Charlotte Haldane (introduced as the 'former wife of Professor Haldane') to draw out the wider political implications of Lysenko under the title of an article which read 'Mine is a moral objection to Moscow'. In this she described Communism as 'a way of living, thinking and working on the intellectual plane, irreconcilable with the moral values of Western civilisation which are based on the moral and religious principles of Christianity and on the ethical and scientific disciplines of the ancient Greek philosophers such as Aristotle and Socrates'.[42]

Haldane's difficulties were also seized on by his old antagonists in the Eugenics Society. In the 1930s he had engaged in a number of intellectual skirmishes with them and in 1946 the battle was joined again. At this time the eugenics movement was on the defensive. The revelation of the Nazi eugenic experiments put them in an extremely embarrassing position. In the October 1946 issue of the *Eugenics Review*, there was an attack by C.P. Blacker on Haldane which reflected the eugenics movement's sensitivity on this issue.[43]

The Lysenko episode came as something of a relief and the *Eugenics Review* went on the counter-attack against Haldane. 'As we go to press,' the editor wrote in January 1949, 'we have been reminded that the facts of this sad story (Lysenko) were first made known to the public by Dr J.R. Baker a fellow of the Eugenics Society.'[44] The attack was then taken up against Haldane.

> A few weeks ago Mr A.J. Cummings of the *News Chronicle* invited Professor Haldane to state his views on the 'new genetics' and their political background and consequences. Professor Haldane replied in the *Daily Worker* with an article that was brilliant debating and thus excellent entertainment. He agreed with Lysenko in this, he disagreed in that. How intellectually refreshing it must have seemed to his readers; what incontrovertible proof of the complete freedom of discussion within the Communist Party. As for Vavilov Professor Haldane did not commit the indelicacy of dragging his name

in, and for the *Daily Worker* we may safely assume that out of sight was out of mind.[45]

The Society for Freedom in Science also subjected Haldane to mounting criticism. J.R. Baker attacked him in a pamphlet issued by the SFS and Sir Henry Dale, in a Foreword to the account of the controversy by the journalist John Langdon Davies, talked of the doctrines of Lysenkoism as the 'nonsense which, with the notable exception of Professor Haldane, the world's geneticists find in them'.[46] The remainder of the book was even more virulent against him. 'Since the decree of 26th August', wrote Langdon Davies in reference to Haldane, 'no human being can both be an honest scientist and a Communist.'[47]

As for the popular press there was almost total silence. This probably reflected a certain haziness in the minds of the leader-writers about heredity. Certainly their occasional excursions into the Lysenko controversy concentrated on the politics of it. The *Daily Mail*, for example, in an editorial on 28th August 1948 claimed the Lysenko affair 'reveals a twentieth-century state is reverting to medieval methods in deference to medieval minds . . . And yet it is understandable in a country where a research worker may be shot for producing a result which offends against the Big Boss. In the West science is intellectually the purest of pursuits. The private beliefs of the investigator count for nothing and the results of experiments are unrelated to politics.' So far so good. But the *Mail* stepped hesitantly into the thorny question of the relationship between heredity and history. 'Communism teaches that all men are equal which means that no man can inherit better qualities than another. He can only progress through the conditions in which he lives.' Reason, according to the *Mail*, always overcame the forces of darkness. But here the leader-writer's confidence gave out and seeing the question of heredity and history slipping from his grasp he brought his column rapidly to a close, unfortunately by something verging on Lamarckianism. 'Rome, it is true, fell to the Barbarian but only because her blow had become enervated by malaria and her vitality undermined by the policy of "bread and circuses".'

One exception to this silence was the *Daily Worker* itself for which Haldane wrote a regular scientific column. The issue of Lysenko emerged in Haldane's column on 6th October 1948 in an article on genetics and evolution in which he referred to the fact that 'some leading biologists in the Soviet Union believe in the hereditary transmission of acquired characters in plants.' Throughout 1948, 1949 and 1950 Haldane continued to contribute articles for the *Worker* a large proportion of which touched on issues of heredity and were exegeses of Mendelian principles. If the readers of the *Worker* in this period took any steps to acquire a biological education from its pages, it was on the whole a firmly anti-Lysenkoist one.

The *Daily Worker* did not switch lines or prevent Haldane from continuing to write in its columns. It did however pursue a policy of giving space to the pro-Lysenkoists and publishing their texts. This led to a sharpening of the antagonism between the pro- and anti-Lysenkoists which was expressed in the pages of the *Worker*. Clemens Dutt on 7th November 1948 referred to Lysenko as expressing 'in sharp form the unsatisfactory nature of the basic theory of bourgeois genetics'. Haldane produced an indirect 'reply' to this on 13th December which was the strongest condemnation of Lysenko and defence of genetics which had appeared till then in the pages of the *Worker*. Haldane admitted that mis-statements by geneticists had added fuel to racism but he turned the tables on his critics by pointing out the reactionary character of the theory of inheritance of acquired characteristics, a point which a reader's letter also made. According to this, 'We have been pushed into defending Lysenko against people like Darlington and Fisher who want only to engage in hysterical tirades against Soviet Science when what we ought to have been doing is defending Soviet Science in general. Incidentally if Lysenko's views are true and applicable to man . . . Families that have ruled for generations ought to go on ruling.'[48]

However the Communist Party made a number of efforts to secure a favourable reaction to Lysenko. A series of meetings of Party scientists under the auspices of the Engels Society were held but these were too divided to come to any firm conclusion. According to one observer the pro-Lysenkoites were generally in a minority.[49] But in spite of their minority position the pro-Lysenkoists were given considerable space to defend the 'new' science. Page Arnott wrote a rather bad demagogic defence of Lysenko in the *Labour Monthly*.[50] Hyman Levy, in the same magazine, was more circumspect. He suggested that a rapport might eventually be achieved between Lysenkoists and Mendelians. 'If geneticists here would be scientific enough to imagine the possibility that Lysenko's claims to be able to produce new hereditary types through environmental control may be substantiated then what has happened would fit into place automatically without far fetched political explanations.'[51] The Communist Party also published a number of pro-Lysenkoist texts. These came out in 1950 and 1951, however, largely after the issue had been settled in the minds of most scientists one way or another. They were more in the way of *post hoc* justification than contributions to the debate. Faced with a considerable amount of resistance the Communist Party was never able to force a line on Lysenko. The maximum it was able to do was to give space to those who were prepared to defend him. This alone was sufficient to link the fortunes of Lysenkoism with the British party and alienate a number of scientists. The final departure of Haldane from the Communist Party in the early 1950s underlined this.

Haldane had set out his views on biology and Marxism in 1948 shortly before it became necessary to confront the question of Lysenko. In 1949, however, Bernal and Haldane debated the issues raised by Lysenkoism in the pages of *Modern Quarterly* with Bernal taking a position favourable to Lysenko and critical of genetics. In his article, Bernal drew the reader's attention to a number of facets of conventional genetics. Genetics, he claimed, had treated creation as the 'isolation and autonomy of the gene . . . in which the whole of creation is determined in advance, being merely an unrolling, literally, an evolution of a plan determined at the beginning of time'.[52] It had attributed to the gene a quasi-metaphysical character, an unchanging essence which stamped itself on life. Lysenko's willingness to accept chromosomes, which he could see, and reject genes — which he could not — was attributed by Bernal to a certain materiality in Lysenko's philosophy which brought it close to Marxism. In addition, according to Bernal, evolution by random selection — a hypothesis upon which population genetics based itself — implied a religious attitude to life. It denied causal connection. For Bernal classical genetics was 'bourgeois' genetics.

This notion fitted with his historical interpretation of scientific developments. According to Bernal the organisation of science and even its theoretical character would be transformed by the emergence of a socialist society. Lysenkoism might very well be considered to fit the role of the new 'proletarian' science. Certain aspects of it, the emphasis on 'practice' and the possibility of a rapid evolutionary transformation of species seemed particularly apposite for a 'socialist' biology.

Haldane, in reply, admitted that the way in which genes had been talked about by geneticists attributed to them a kind of metaphysical character. But he went on to argue that genes were not transmitted independently of the organism or environment.

That they were not immutable and that the expression of the potentiality in genes was dependent on environmental influence. All of these points, Haldane argued, met Lysenko's objections to genetics. Moreover geneticists had adjusted their work to incorporate these ideas. Haldane implied that the scientific practice of geneticists was 'material' even if the language in which it was expressed was not. The logic of scientific development was to force an increasing recognition among scientists of the materiality of living systems whatever 'idealist' notions might be residual in their work. As for random variation, Haldane argued that it did not imply the absence of causality. It served only as a hypothesis upon which to build another theory. The 'cause' of genetic variation was not ruled out *a priori* from investigation even though a theory of evolution had been constructed without a full understanding of it.

Lysenko's claim to have improved yields in Soviet agriculture was frequently cited by his supporters. Haldane remained sceptical.

> Lysenko and his supporters point with justifiable pride to the very great increase in productivity which has occurred in many parts of their country in recent years. If I did not regard its economic system as superior to that of my own country, I should be forced to suppose that its methods of livestock improvement were greatly superior. But I am much more convinced that collective farming is superior to capitalist farming than that Soviet breeding practice excels our own.[53]

As for the question of how genetics had come to be associated with right-wing politics whereas Bernal believed that, 'The connection of orthodox genetics with eugenics with Malthusianism and with theories of race superiority and ultimately with Nazism are not accidental. Its foundations need to be closely examined, criticised and amended before they can be admitted safe to build on.'[54] Haldane stated his case in this way:

> I am a Darwinist although Darwin (1879) wrote 'Man, like every other animal has no doubt advanced to his present condition through a struggle for existence consequent on his rapid multiplication, and if he is to advance still higher, it is to be feared that he must remain subject to a severe struggle. Otherwise he would sink to indolence and the more gifted men would not be more successful in the battle for life than the less gifted.' Similarly I am a Mendelist-Morganist, although

Mendel used an idealistic terminology and Morgan wrote of the mechanism of heredity. But Morgan and his colleagues made the very great advance of showing that heredity has a material not a meta-physical basis. Their discovery underwent the normal fate of all advances towards materialism. It was mechanistically interpreted. And it is often taught in a manner which combined mechanism and idealism. Thus geneticists sometimes say that an animal has the same gene as its father in a given locus as if genes combined the property of indestructibility with the still more remarkable one of being in two places at once. But that does not mean that we should reject the large element of genuine, constructive materialism in Morgan's views.[55]

Alan Morton in *Soviet Genetics* (1951) set out what was in fact a summary of Bernal's views on the social determination of science. Even whilst he argued that the question of Lysenko was decided by scientific criteria alone, in common with other pro-Lysenkoists, he emphasised the division of science into bourgeois and proletarian. Science reflected capitalist exploitation in its organisation and application but also 'these features can also reflect themselves in a more subtle way in the ideas, the theory of science.'[56] Mendelism, he argued, exemplified this.

In contrast, Haldane's view seemed to approach a Marxist framework only hesitantly. Haldane, in his writings on Lysenko, seemed to suggest a number of general propositions about the relationship between science and society. Briefly, they could be summarised as these: the history of each science has to be constructed for that science alone. Thus whereas Bernal, Morton etc. talked in omnibus categories (and indeed their intervention within the controversy was generally justified by a general theory of science and its proletarian or bourgeois character), Haldane struggles with the actual history of Mendelism. There were certain advances in this theory but also mechanical or idealist interpretations. Secondly, the ideological elements in these he associated with certain philosophical interventions — with idealism and mechanism — just as Bernal did. But his reflections on this suggested that these idealist elements could co-exist with genuine scientific advance — a point which he merely stated without explaining. Lastly he seemed to be saying that part of the problem lay in the language in which biological science was expressed — again a problem specific to that science. Haldane made no general claims beyond the area of biology.

These points were relatively under-theorised in contrast with Bernal's. Thus Haldane's justification of his position looked like pragmatic concessions to both sides. Moreover Bernal's theory allowed a number of neat 'moral' criteria to be applied. It reproduced a theoretical division that was meant to reflect the political division of the world and could act as a rallying cry for one side or the other. All Haldane could offer was hesitation, approximation and the possibility of an explanation of a small area of science. In these circumstances Bernal's views exercised more persuasion than Haldane's. Both right and left looked for a simple reflection of social ideas in science, Polanyi as much as Bernal, as we shall see. Bernal's demonstration of the connection between Lysenkoism and socialism appeared more convincing than Haldane's rebuttal of it. This led some scientists to support the 'new genetics'. It led many more to say, 'so much the worse for Marxism'.

The scientists without specific commitment to Marxism approached the problem in a different way. Both Waddington and Huxley wrote on the controversy. In the *New Statesman*, Waddington attacked Lysenko's scientific principles but he baulked at the attack on political control of science which Lysenko's antagonists had engaged in. 'Political action based on the criticism of current scientific theories cannot be considered as an isolated phenomenon and condemned because there is something special about science which exempts it from control; such action is part of a general social philosophy and can only be morally judged as such.'[57] Huxley on the contrary had moved to a position where this was precisely the problem. 'All these issues are, I repeat, either irrelevant or merely subsidiary to the major issue which is the official condemnation of scientific results on other than scientific grounds, and therefore the repudiation by the USSR of the concept of scientific method and scientific activity held by the great majority of men of science elsewhere.'[58]

The recounting of these opinions does not go far enough in providing a picture of the divisive impact of these events. Those scientists committed to Marxism were in disarray. Many liberal scientists were obviously alienated by the events in the USSR. Moreover the Lysenko controversy allowed the SFS to exercise a greater influence than they had ever done. It allowed them in fact to go onto the counter-attack. All organisations of scientists found it increasingly difficult in this climate to maintain an effective independent opinion. This was not simply due to Lysenko. A whole series of developments produced tensions between the government and the scientific community but — as in the case of atomic power — Lysenko played a role in sapping concerted opposition to certain policies.

Notes

1. Z.A. Medvedev, *The rise and fall of T.D. Lysenko* (Columbia University Press, New York, 1969); B. Lewonton and R. Levins, 'The problems of Lysenkoism' in *The radicalisation of science*, H. Rose and S. Rose (eds), (Macmillan, London, 1976); D. Joravsky, *The Lysenko affair*, (Harvard University Press, Cambridge, Mass, 1970); D. Lecourt, *Proletarian science?* (New Left Books, London, 1977).
2. 'Extinction of human rights in Eastern Europe', PRO Cabinet Papers, CP (47) 313.
3. Eric Ashby, 'Genetics in the USSR', *Nature*, vol. 158, 31 August 1946, p. 287, (review of P.S. Hudson and R.H. Richens, *The new genetics in the Soviet Union*, Cambridge, 1946). For an explanation of Lysenkoism, see Introduction of this book.
4. K. Sax, 'Soviet science and political philosophy', *Scientific Monthly*, vol. 65 (1947), pp. 43-7.
5. T. Dobzhansky, 'Lysenko's genetics', *Journal of Heredity*, vol. 37, no. 5 (1946), p. 5.
6. Julian Huxley, 'Evolutionary biology and related subjects', *Nature*, vol. 156 (September 1945), pp. 254-6.
7. J.R. Baker and A.G. Tansley, 'The course of the controversy on freedom in science', *Nature*, vol. 158 (26 October 1946), p. 575.
8. Ibid.
9. I.M. Polyakov, 'Conference on genetics and selection', Moscow (7-14 October 1939); reprinted in *Science and Society*, vol. 4 (1940), pp. 183-437, p. 232.
10. Ibid., p. 436.
11. J.L. Fyfe, 'The Soviet genetics controversy', *Modern Quarterly*, vol. 2 (1947), pp. 347-51.
12. John Lewis, 'A footnote on the Soviet genetics controversy', *Modern Quarterly*, vol. 2 (1947), pp. 352-6.
13. George Bernard Shaw, 'The Lysenko muddle', *Labour Monthly*, vol. 3 (January 1949), pp. 18-20.
14. J.B.S. Haldane, 'Biology and Marxism', *The Modern Quarterly*, vol. 3 (1948), pp. 2-11.
15. John R. Baker, Letter of 2 November 1940; reproduced in Publications of the Society for Freedom in Science (1953). The founding members of the Society were: president, Sir George Thomson; vice presidents, Sir Henry Dale, Professor Michael Polanyi and Sir Arthur Tansley; members, Professor Blackman, Dr Sutton and Dr Sherwood Taylor; representative for the USA, Professor P.W. Bridgman; honorary secretary and treasurer, Dr J.R. Baker.
16. Ibid.

17. Ibid. The science journalists he referred to were probably J.G. Crowther (*Manchester Guardian*) and Ritchie Calder (*News Chronicle*).
18. Baker. Publications of the Society for Freedom in Science (1953).
19. Michael Polanyi, 'The autonomy of science', *Scientific Monthly*, vol. 60 (1945), pp. 141-50; reprinted from 'Memoirs and proceedings of the Manchester Literary and Philosophical Society' (Session 1941-3).
20. Baker, Publications of the SFS, p. 10.
21. Ibid.
22. C.D. Darlington, 'Retreat from science in Soviet Russia', *Nineteenth Century and After*, vol. 142 (October 1947), pp. 157-68.
23. Ibid., p. 159.
24. Ibid.
25. An example of this is C. Zirkle, *Evolution, Marxian biology and the social scene*, (University of Pennsylvania Press, Philadelphia, 1949).
26. Reprinted from the BBC Third Programme in *The Listener* (4 November 1948), pp. 677-8.
27. *Listener* (9 December 1948), pp. 873-5.
28. *Atomic Scientists News*, vol. 2, no. 6 (28 April 1949), p. 137; and *Daily Worker* (7 May 1949).
29. *Manchester Guardian* (2 September 1948).
30. *Manchester Guardian* (16 September 1948).
31. Ashby, *The Listener* (4 November 1948), p. 677.
32. Ibid., p. 678.
33. R.A. Fisher, *The Listener* (9 December 1948), p. 874.
34. The resignation and letter were reported in *The Times* (26 November 1948), and in other newspapers. The full text was reprinted in *Discovery*, vol. 10 (January 1949), p. 32.
35. *The Times* (1 December 1948).
36. Ibid.
37. Herbert Morrison, *The Times*, 1 December 1948.
38. Ibid.
39. Ibid.
40. A.J. Cummings, *News Chronicle* (1 October 1948).
41. Ritchie Calder, 'Seeds of strife', *News Chronicle* (29 November 1948). According to Calder, 'What we want to hear from Haldane, whose statistical genetics have been denounced, is a rational, scientific defence or disclaimer of practices which (apparently) strike at his life's work. And at a great deal more.'
42. Charlotte Haldane, 'Mine is a moral objection to Moscow', *News Chronicle* (22 December 1948).
43. C.P. Blacker, 'Social problem families in the limelight', *Eugenics Review*, vol. 38, no. 3 (October 1946), p. 117.
44. C.P. Blacker, *Eugenics Review*, vol. 40, no. 4 (January 1949), p. 177, 6n.
45. Ibid., p. 177.
46. John Langdon Davies, *Russia puts the clock back* (Gollancz, London, 1949) foreword by Sir Henry Dale, p. 7.
47. Ibid., p. 77.
48. Letter from Anthony Barnett, *Daily Worker* (7 December 1948), p. 2.
49. An account of this episode can be found in G.A. Almond (ed.),

The appeals of Communism (Princeton University Press, Princeton, 1954); and in Henry Pelling, *The British Communist Party* (Adams and Charles Black, London 1958).

50. Page Arnott, 'Scientists in livery', *Labour Monthly* (February 1949), pp. 55-60.

51. 'The Soviet controversy', Hyman Levy, *Labour Monthly*, (April 1949), p. 119.

52. J.D. Bernal, 'The biological controversy in the Soviet Union', *Modern Quarterly*, vol. 4 (1949), pp. 203-17, p. 206.

53. J.B.S. Haldane, 'In defence of genetics', ibid., pp. 194-201.

54. Bernal, 'The biological controversy', ibid., p. 213.

55. Haldane, 'In defence of genetics', ibid., p. 198-9.

56. Alan Morton, *Soviet genetics* (Lawrence and Wishart, London, 1951).

57. C.H. Waddington, 'Lysenko and the scientists' (part 1), *New Statesman*, (25 December 1948), p. 566; part 2 appeared in the *New Statesman* on 1 January 1949, pp. 6-7.

58. Julian Huxley, 'Soviet genetics: the real issue', *Nature*, vol. 163, (18 June 1949), pp. 935-42, p. 935.

3

Into Two Camps

In the American *Bulletin of Atomic Scientists*, Lysenko was given considerable coverage. According to H.J. Muller, the task of all scientists regardless of their particular specialism was in 'checking the already dangerous spread of the present infection to countries outside the Soviet Hemisphere and of making clear to the people of these countries the important lessons for culture and for civilisation in general which are involved'.[1] In 1949, a complete issue of the Bulletin was devoted to Lysenko. According to the editor, 'It may seem strange that the *Bulletin of Atomic Scientists* should devote a large part of an issue to a review of events in the field of genetics which occurred in Russia almost a year ago . . . We state here these lessons of the purge not as our contribution to the "cold war" but to encourage a long perspective as to the consequences of "statism" for the free growth of science.'[2]

The question of atomic policy and the atomic arms race was a crucial issue for the scientific community in the 1940s. The pressure groups of scientists which had emerged out of the radical 30s tended to regard the need for restriction of the political exploitation of atomic power as a legitimate concern of scientists as a group. Further they were more sympathetic to a form of international supervision which, at that time, would have made the American monopoly of atomic weaponry less effective. An American writing in *Science and Society* in 1947 pointed out the successes he claimed scientific pressure groups had achieved in the area of atomic policy. The Federation of American Scientists had publicly criticised the Bikini explosion for its military rather than scientific purposes. They had, he claimed, prevented the passing of the 'May-Johnson' bill and the Vandeburg amendment to the MacMahon bill (which was, however, passed later). They had also prevented the passing of the Magnusson-Kilgore National Science Foundation bill.³

In Britain the stakes were less for, whilst the Americans had already developed the atomic bomb, the British had to wait until 1952 before they successfully tested one. None the less the British government decided in 1945 to embark upon a nuclear weapons test programme. Blackett was on the scientific advisory committee to advise the British government on atomic questions. The other members of the advisory committee included Sir Edward Appleton, Sir Henry Dale, Sir George Thomson, Sir Robert Robinson, Sir Henry Tizard, Sir James Chadwick. In 1945 Blackett presented a paper before the committee arguing that Britain should not produce its own atomic bomb.⁴ The chiefs of staff and the Prime Minister disagreed. In February 1947 Blackett produced another memorandum on similar lines. The Foreign Office called his views 'dangerous and misleading rubbish'. Ernest Bevin pencilled in the margin 'He ought to stick to science.'⁵ In 1947 the advisory committee lapsed and from then on decisions on atomic weapons were taken by a small group of cabinet members whose deliberations were kept from Parliament and the other members of the cabinet.

Blackett, however, continued to publicise his views. In 1948 he published *Military and political consequences of atomic power*. Edward Shils, of the University of Chicago and in the late 1940s reader in social theory at the London School of Economics, attacked him in the pages of the *Manchester Guardian*. Shils described Blackett as writing 'from the Stalinist viewpoint: all his main conclusions and premises are to be found in the pamphlets of ordinary Communist propagandists, the "conspirational theory of society", the alleged yearning of America for a "preventative war" against the Soviet Union: the good and virtuous ground of every Soviet action, however offensive it feels at first sight.'⁶

There were two organisations of scientists which concerned themselves with atomic questions, one was the AScW and the other the Atomic Scientists Association (ASA). The ASA was founded in 1946. It included many of the scientists who had been involved with the development of atomic physics in the 1930s and the subsequent development of the atomic bomb. Some of them had worked on the Manhattan project at Los Alamos. It also included scientists who were still deeply involved in the British atomic bomb project as well as those who had since retired into academia. The range of political opinion in the ASA was wide. There were radicals and staunch conservatives among them.

Its president, from 1948 to 1950, was Rudolf Peierls who, with

Otto Frisch, had presented the British government with a memorandum in 1940 stating the logistics of atomic bomb construction. The Peierls-Frisch memorandum was an important stepping stone on the route to an atomic bomb and was particularly important in the development of the British atomic bomb project. Subsequently Peierls had worked at Los Alamos with Klaus Fuchs, like him a refugee from Nazi Germany, as his assistant. In 1945 he returned to Britain as Professor of Mathematical Physics at Birmingham University.

Peierls, like many other scientists, was struck by the moral implications of nuclear weapons. In 1946 he took the time to write to a young correspondent who had read an article of his in *Reynolds News* on the question of the morality of force in politics.

> I also agree with you that there is little difference between the destruction of civilian life by our bombs in Germany or by atomic bombs in Japan and that of neutral or allied civilians by the Germans. It is a tragedy of the war that, in trying to wipe out the powers responsible for such crimes we had the choice between committing the same crimes or perishing ourselves. I also agree with you that much of the actions for which western civilisation is responsible such as the fate of Red Indians and Australian natives (or might I add here the treatment of negroes in America and South Africa) differs from what the Nazis did only in degree.[7]

Peierls went on to outline why, in these circumstances, he would still condone the use of force by the Allies. He believed that, unlike the Nazis, the allies did not glorify brutality and were capable of learning. But he conceded that this moral superiority was relative and fragile to a degree.

Peierls took the view that the influence of scientists should be exercised on governments to persuade them of the importance of some kind of international control of atomic weapons. He thought the newly formed ASA could play this role, although he was aware that politicians were rather reluctant, even in the more optimistic days of 1946, to receive advice from scientists on these matters. Peierls wrote to Sir John Anderson, then chairman of the advisory committee on atomic energy (1945-8), in charge of trying to construct bridges between the ASA and prominent politicians:

> When I last saw you in Washington you expressed anxiety

that scientists should, in stating their views publicly, bear in mind that they may embarrass the statesmen who, fundamentally are trying to achieve the same objects, but whose actions are, of course, limited by the political realities.

Since then the scientists have formed an 'Atomic Scientists Association' run at present by a Provisional Committee of which I am a member. I am anxious to assure you that in forming such an organisation we are bearing in mind the argument which you stressed and that our desire is to be constructive and to do what we can to be helpful to the endeavours to find a reasonable solution to the serious problems of the fate of atomic energy.[8]

On 12 December 1947, the executive committee of the AScW wrote to *The Times* arguing that, 'We feel strongly that our government should take a stand of greater independence of the American point of view on the subject (atomic weapons control) and should strive to effect a compromise between America and the USSR.'[9] The AScW in a memorandum in August 1947 had set out its views on how this compromise could be achieved.[10] Similarly the ASA issued its own statement in January 1947. This statement was largely on the same lines as that of the AScW but more conservative in what it considered could be achieved. It also required less movement on the part of the Americans than the AScW document. Much to the embarrassment of the ASA however, it found itself lumped together with the AScW and quoted by the Soviet representatives, Vyshinsky and Molotov, in speeches at the United Nations. The members of the ASA council complained:

> The biased selection of the quotations, together with the fact that in Vyshinsky's speech the views expressed by our Memorandum are not clearly separated from those given by the AScW Memorandum may create a false impression of where we stand, and from the press reports many people might have got the impression that our Association is dissatisfied with the US and British stand on international control and supports the Soviet view.[11]

To clear up this the ASA issued a statement pointing out what they considered to be misrepresentations of their views. In fact their views, in the context of British and American negotiations, were controversial. But the ASA realised that to be tarred with pro-

Sovietism was highly damaging to any attempt to push their government towards international control of atomic weapons. None the less the ASA continued to speak out on the subject. In 1948 the ASA followed with a new memorandum on the subject. This argued for some kind of central control over the production of atomic weapons, for parity between the great powers and for an improved political climate between East and West. It also called for international cooperation between scientists.

Two years, however, had made a considerable difference. Whilst the British government was prepared to hear and pay attention to the AScW and the ASA on questions such as the Atomic Energy Bill and on nuclear safety, it was not prepared to listen to representations on the political control of nuclear weapons. In addition press comment was largely unfavourable. *The Economist*, commenting on the memorandum on 24th July 1948 claimed, 'It is difficult to see how anything could come of such a "collaboration" except a one-way traffic to the disadvantage of the Western democracies.'[12] *Nature* in an editorial on 14th August denounced the proposals in similar terms. According to the ASA this was an event of particular importance since *Nature* 'has always stressed the international aspects of science and it shows strikingly the way in which military considerations can affect the outlook of scientific workers and lead them to adopt against their will a totalitarian view of their function.'[13]

These two years not only saw the deterioration of the international situation but also increasing pressure on atomic scientists particularly on the question of security. In 1946 Alan Nunn May had been tried and sentenced to ten years penal servitude for passing atomic secrets to the USSR. In 1949 Fuchs was arrested and charged with passing atomic secrets to the USSR, a charge which he admitted. In 1950 Fuchs himself was given a sentence of 14 years. In 1950 Bruno Pontecorvo, the Italian physicist working at Harwell, defected. In the USA the trials of the Rosenbergs followed in 1951.

Apart from the wave of hysteria the existence of atomic spies produced in the USA, the British government was particularly disturbed. It hoped that the American government might be persuaded to resume the co-operation over the nuclear arms programme which had existed during the war. But these events deeply embarrassed the British governments and made it decreasingly likely that the USA would allow a free interchange of atomic information with Britain. The spy revelations fed anti-Communism in both Britain and the USA, for the motive of those involved in spying for the USSR was political sympathy with the Soviet Union. For

British scientists the blows were particularly painful for not only had the spies worked among them, but also there was residual sympathy for the idea of free exchange of information internationally and some dislike of the attempt to impose strict barriers against it. Several prominent scientists petitioned the government in mitigation of Nunn May's offence on these grounds.

For Peierls the arrest of Fuchs was a shattering personal blow. Peierls drew up a statement on the case of Fuchs pondering its significance. He had appointed Fuchs knowing he was a left-winger but in the 1930s it was the left who had opposed Hitler. Fuchs had been approved by the authorities. In conversation he had been left-wing but 'scornful of dogmatic Communism' and had hidden his beliefs.[14] Peierls' statement put its finger on a real fear. The case, he believed, could well lead to a campaign against those who had been left-wingers in their youth, against foreigners and against scientists.

This was a prescient view. A peculiarly nasty and sometimes sinister campaign was carried on in the press against all three categories. The *Evening Standard* on 22nd December 1947 carried a particularly virulent piece by the novelist Rebecca West about Nunn May, 'His betrayal may harm us all'. The tone was captured in this passage. 'It had been the claim of the violent men who formed the Nazi-Fascist movement that they were endowed with a greater amount of physical strength and vitality than the mass of the population; an amount which would enable them to seize that power if it were denied them. It was now the claim of the Communist-Fascist movement that they were endowed with a greater amount of special technical knowledge than the mass of the population, an amount which would enable them to seize that power if they were denied it.' The piece also contained exaggerated claims about Nunn May's role in Soviet atomic weapons' development and veiled accusations against scientists in general. The physicist N. Kurti cut it out and sent it to Peierls with the comment, 'it made me almost sick.'[15]

With the arrest of Fuchs and the defection of Pontecorvo, an anti-foreigner tone began to invade sections of the British press. Peierls took particular exception to an article on 'foreign-born' scientists, dealing with the Pontecorvo case and published in the *Sunday Express* in 1950. In that article the names of Kurti, Frisch, Born, Rotblat, Simon and Peierls himself — all perfectly innocent — were mentioned. Peierls was provoked to write to Cudlipp the editor:

No doubt the results of such cases must be to set people wondering about those who remain and to make further enquiries, but I was always sure and I am still sure that this will be done in a spirit of fairness, and that those who have nothing to hide have nothing to fear.

But what if I am wrong? That would be a disaster, not for myself but for the world. It would mean the loss of the greatest tradition of justice and fairness and respect for human rights.[16]

Peierls had a clear idea of the consequences of these events. He could not join in the political suspicion and xenophobia which overtook some quarters of British life. However, he felt that certain freedoms would, without overt restriction, be difficult to exercise. He wrote to Nevil Mott on the Fuchs affair:

As you may see the Fuchs affair was a dreadful blow. Probably not so bad even for us as for some of the people at Harwell where he had built up a team of young people who had enormous confidence and admiration for him and who knew him (as indeed we knew him) as an unselfish man with a passionate devotion to his work and ready to go to any trouble for his friends and subordinates. However these personal factors pall compared with the effect this case will have on the political atmosphere and on the position of scientists here and America and for relations between the two countries.[17]

As we shall see, these events influenced Peierls' attitude as to what it was politically possible for the ASA to achieve.

There were some in the ASA who felt that it was unwise to enter the arena of arms negotiation with political advice at this point in time and some who had always been hostile to the idea. The latter were also gaining strength. H.W. Skinner, Professor of Atomic Physics in Liverpool, stated their position when in the February before the ASA memorandum of July 1948 he wrote:

I think our future plan of operation should be very carefully thought out. There is, for example, no hope whatever of getting agreement on 'The consequences of this country having a long range striking force equipped with atomic bombs' . . . we are in a dilemma and, no matter how we try to cover things up, the alternatives plainly remain a) to become a professional association b) to become a political association. The first alternative seems best. We would then give expression to a wide variety of opinion . . . but would not try to have any official opinion on debatable subjects.[18]

None the less a statement on arms control was issued by the ASA in July and discussions were scheduled for the ASA conference in October. Whilst this was of only marginal interest to many scientists, it was an additional problem for the left-wing scientist who wished to ameliorate the damaging split which had grown up between East and West.

At this point the Lysenko controversy erupted. At the Conference of Atomic Scientists in October 1948, Sir Henry Dale argued against contact between British and Soviet scientists, using as justification the Lysenko controversy. As reported in the Bulletin of the ASA:

Sir Henry denounced in very strong terms the coercion of geneticists in Russia. 'Unless a geneticist will profess and proclaim belief in what he, with every responsible geneticist in the world knows to be nonsense, he is dismissed and liquidated. There has been nothing like it since an earlier orthodoxy sent Giordano Bruno to the stake and the cardinals bullied Galileo who disclaimed what he knew to be scientific truth.' The speaker added that although he was in favour of the widest possible cooperation between scientists of all countries, we must be careful that, in attempting to attain this end we do not inadvertently give countenance to such a state controlled form of pseudo science and thereby discourage our suffering colleagues in this country.[19]

Sir Henry Dale, however, met some opposition. Dr Burhop, for example, objected to Sir Henry Dale's reference to the liquidation of scientists. In the face of this, Sir Henry Dale backed down. As reported in *Atomic Scientists News*, he said: ' "Perhaps I should not have used the word liquidate with a general application but he referred to Vavilov." However he admitted that he did not know whether this could be applied to Vavilov but claimed his death was in a shroud of secrecy which the Russians would not lift'.[20]

The acrimony of this debate meant that no conclusive position on atomic weapons was taken by the conference. Indeed as E.S. Burhop, Professor of Physics at University College London and a strong supporter of the left, wrote to Blackett 'In a session supposed to discuss atomic energy and the scientists, most of the time was spent in debating Soviet genetics.'[21] Lysenko, in fact, became one more issue dividing scientists and putting the right on the attack, the left on the defensive and demoralising the liberal middle of the political spectrum.

The divisions within the ASA multiplied. There were serious disagreements over whether the ASA should protest at the Civil Service purge of scientists in 1948. In the end a statement was issued condemning it. In spite of protests, Peierls insisted that the work must go out:

> It has taken us a long time to say anything because there was such a strong division among council members between those who would not have any political discrimination of any kind and those (including myself) who felt that one could not oppose such security measures but one would watch that they did not interfere with efficiency and that they did not spread to other places where there is no secret work going on.[22]

In October 1949 after the explosion of Soviet Union's atomic bomb, the executive committee stated that, 'Because of these divisions of opinion no general statement on the issue representative of the view of the majority of the members can be made.'[23] Council also disagreed on reaction to the USA's decision to proceed with an H-bomb project in 1950 and the press statement that was drawn up could not be issued as a statement of the ASA itself due to these disagreements. In 1951 a further controversy broke out. A civil defence exhibition was set up at which the ASA was asked to contribute a stand. Because the ASA's contribution laid stress on the destructive capacity of the atomic bomb and refused to allow propaganda for civil defence recruitment, Lord Cherwell turned his wrath on Peierls. It was, he believed, after reading the council minutes, 'So unpatriotic that I would have assumed some typing error'.[24] Peierls wrote back defending the decision:

> The point is that in the view of many members of Council there is considerable doubt whether the present civil defence programme, as far as it relates to atomic bombs is being planned in a sensible and realistic way. We do not know enough about it to form a clear opinion, but we are disturbed by the many reassuring statements which claim that

preparations on the scale now considered can considerably reduce the scale of the disaster caused by an atomic bomb.[25]

The division of opinion in the ASA gradually led to its depoliticisation. By 1951, there were increasing calls for the ASA to stand aside from politics altogether. H.W.B. Skinner wrote in to state that, 'There seems to be a tendentious idea in some scientific minds that a "scientist" as such has something special to contribute to the solution of world problems in general apart from his specialised technical knowledge. I do not believe that there is any evidence this is the case . . . I should therefore like to propose that the Association should abstain altogether from political subjects.'[26] George Thomson — SFS member and former member of the Government's Advisory Committee on Atomic matters — agreed: 'If we can be factual we have a useful future. If we insist on trying to be political we have none.'[27] Lord Cherwell agreed. 'Bodies like the Royal Society did not indulge in politics; why should we?'[28] The Council of the ASA agreed with him though Dr Kurti and Professor N.F. Mott disagreed. Members of the ASA like N.F. Mott, Kathleen Lonsdale and Dr E.H.S. Burhop stood out against this trend but they were increasingly isolated. By 1951 the Bulletin of the ASA had converted itself into an almost wholly academic journal whose editorial comments largely reflected conventional government opinion on atomic policy. Moreover it could not be relied on even to fight the restrictions on personal freedom placed on its members. For example Dr E.H.S. Burhop was informed in September 1951 that his passport had been cancelled. A ban on travel to the USSR affected Kathleen Lonsdale, a Quaker and a member of the ASA. Protests were only perfunctory.

Some members of the ASA were disturbed by this turn of events. J. Champion, a member of the ASA's council complained about the ASA's increasing tendency to duck contentious issues, particularly over atomic weapons:

> Not unconnected with these difficulties is the question whether our Association can continue to function. How prominent members of our Association can still be expected to have a disinterested scientific opinion on vital issues, when they are already tied to existing authority as government consultants, official civil servants, by the Official Secrets Act is a difficulty I cannot resolve. Massey and others agree that this issue of the continuance of our Association depends on whether or

not we have anything to say on the H-bomb. I agree it was easy enough for the ASA and governments to see eye to eye in the deceptively optimistic atmosphere of 1945-6 because all were apparently meeting in a constructive direction. I deny that an atomic arms race is a constructive proposition and since the published aim of our Association is to help influence government policy towards a constructive use of atomic energy the least duty some of us can perform to our members is to show how clearly we are aware of the discrepancy.[29]

In spite of misgivings, the ASA did not disappear but by the 1950s it had retreated from the attempt to influence statesmen in the manner Peierls had hoped for in 1946.

The strains produced by Lysenko were also felt within the Association of Scientific Workers. On the whole the reviewers for its journal, *The Scientific Worker* tended to take a Lysenkoist position. N.W. Pirie, reviewing *Soviet Genetics* in December 1948, recorded his disagreements with Lysenko's theories but felt that, 'We cannot dispute the fact that orthodox genetics is making little headway in this direction nor the right of the state to say that much of the research for which it pays should be directed towards its end.'[30] Angus Bateman reviewing 'The situation in biological science, a record of the proceedings of the Lenin Academy of Agricultural Sciences', maintained that 'if all scientific issues were submitted to the same frank and searching debate, the progress of science would be very much promoted.'[31]

The sympathetic reception on the part of the reviewers was not, on the whole, shared by the membership or at least those who chose to write in on the subject. These reviews produced a correspondence which stretched from January 1949 to March 1950. The response to them was overwhelmingly hostile (eleven against and one for). Many of these were not only from the right but from radicals and Marxists. One complained that the Lysenkoists were 'Doing precisely what Engels would have deplored and repudiated namely, building science into a dialectical framework and rejecting any scientific thesis which cannot be put into the register of dialectical materialism. This procedure makes a mockery of the efforts of those of us who are today defending Marxism against those who claim it to be just another dogma.'[32] Another repeated a familiar theme of whether the 'AScW should seriously consider whether it is wise to continue alienating liberal minded scientists by parading apologies for Stalinist religious beliefs.'[33] A political battle was being fought out between right and left at branch level and the cause of Lysenko had not helped the consolidation of radicals within the AScW. To defend certain general lines of policy on conditions and wages and even to put forward a line on foreign policy questions which touched on science was one thing but the defence of what many scientists, including Marxists, felt to be an intellectual aberration was another.

Shortly after the first review the demand that the AScW should disassociate itself from Communism grew more insistent. The honorary secretary from the North West Area sent in a resolution originally passed on 16th January 1949 demanding that the Area Committee would not 'sponsor any action on behalf of the Association which will give intentional support to the Communist Party'[34] and asking for support for the recent memorandum circulated by the TUC which called for Communist influence to be stemmed in the Unions. In June 1949 the Executive Committee issued a statement publicly disclaiming any such association and pledging itself to an independent political line.[35]

Lysenko was, of course, not the only thing which hardened the divisions. The AScW was subject to the TUC's drive against Communists in the Trade Union movement. Depoliticisation of the AScW was regarded by many Labour Party members within the organisation as the first step towards a re-politicisation — this time by an official Labour Party affiliation for which some of them pressed. These had every reason to try and stem the influence of the non-Labour Party left. The tightening of political control within science and a sense of their public unpopularity already had created a degree of political apathy among scientists which sapped organisations like the ASA and AScW both of members and of any other than the most routine aims. In this climate Lysenkoism acted as a final signpost on the road to political disillusion.

In the end the AScW executive responded to the furore by appointing a special committee of investigation to report on Lysenkoism. This was published in 1951. In one way it helped divert the issue away from its part in the day to day affairs of the Union, although by 1951 most of the effects of the Lysenko controversy had already been felt. The committee's report came down against Lysenko whilst attempting a degree of detached comment on the question. The committee, in fact, produced a reply which would have thoroughly appeased Haldane. It concluded that:

(1). The facts elucidated by Mendelian genetics are unchallenged.
(2). The theoretical structure attributed to Mendelism by the Michurinists is outmoded, modern Mendelism being free from the defects pointed out by the Michurinists.
(3). Some Mendelians were certainly responsible for idealistic interpretations of their results but these they repudiated.
(4). Mendelism had been and would continue to be of value to practice.
(5). The results obtained by the Michurinists could be explained in terms of Mendelian concepts or extensions of them.
(6). The attributing of heredity properties to material particles — the genes — could not be considered idealist.[36]

Their comment on the politics of the question is also of interest. According to the committee,

> The elements of the Michurinist argument which arouses most interest outside the USSR are the confidence shown by its expounders in the truth of their philosophy and the equation of the philosophical opinions of the adherents of a scientific theory with a corresponding political position. Thus the use of the words 'scholastic' and 'reactionary' in a scientific discussion would be rejected as irrelevant by most non-Soviet scientists. This, and the acknowledged part played in the controversy by non-specialists from farmers to philosophers constitute a major break with scientific practice outside the USSR.,[37]

Gradually membership of the AScW was declining. By 1954 it had fallen to 11,382 from a post-war peak of around 15,000. The losses were particularly heavy in London, perhaps signalling the easy way in which cultural and political reasons for joining the AScW were affected by the adverse political climate. In the more unionised areas of the countries, where membership was seen as a necessary part of industrial and occupational protection, the figures held up better. The report of the Association of Scientific Workers on Lysenko came too late to affect the process of depoliticisation it helped to create among scientists. Lysenkoism was not the only reason but it was one of the causes for the ebbing away of the 'centre' from associations like the AScW and the increasing difficulties bodies like the ASA experienced in producing an opinion on scientific policy independent of the government. The left meanwhile became increasingly isolated within the scientific community.

This was conceded by the scientific representative of the USSR though in a peculiarly obtuse fashion. At the Botanical Congress in Stockholm in 1950 — which, since the Soviets had cut off all relations with geneticists, was the main forum where Lysenkoism could be discussed with Western scientists — the Swedish scientist Ake Gustafsson expressed a great deal of curiosity about the state of biological science in the USSR. He recounted his meeting with the Soviet delegation. The official leader V.N. Soukatchov had in fact produced Mendelian papers in the past and the delegation itself included, he believed, a number of 'silent' Mendelians. But Mendelism was attacked by the Soviets — particularly the work of Blakeslee and Darlington. The Soviet delegation also proudly made the claim that the Lysenko controversy had 'split' the geneticists abroad.[38] Their claim was too modest. They had, in fact, split the scientific community as a whole not just the geneticists.

If this led to a depoliticisation it was of a very particular and unique kind. What happened was that the ideology about what scientific work was became a-political at the very moment when the actual political connections and state intervention in British science were becoming stronger than they had ever been. Margaret Gowing has estimated that the British government's expenditure on the atomic programme in the years 1945–51 was £200 million. Moreover all of this money had been spent with the minimum of parliamentary supervision and next to nothing of press discussion.[39] Even within the Cabinet the decisions were confined to an inner circle and frequently taken without the knowledge of the rest.

The career structure of science became closely bound up with the progress of government control of certain areas of science. From this process emerged the 'scientist-statesman' for whom Morrison had appealed in his speech to the Royal Society and of whom Lord Cherwell was a good example.[40] For scientists of this sort a career in science meant a strong identification with the policies of consecutive governments and even the development of a rhetoric about science which linked the aims of the West with the progress of science. It also meant that 'statism' which in the early 1940s had been the enemy of the more conservative among scientists was redefined. 'Statism' came to mean simply socialism. Other forms of government intervention in science and even restrictions on the expression of political opinion became acceptable, even desirable.

This can be illustrated in the way that the protagonists of the SFS were beginning to modify the objections they made in the early forties to state-controlled direction of research. In 1951, Sir Henry Dale admitted that:

> A growing proportion indeed of modern scientific research could not be done at all without large subventions from such outside sources. And under such conditions, however much we may desire that Government Advisory Councils, Charitable foundations or a benevolent industry should be ready to show their faith in individual scientists by simply endowing them to do whatever research they fancy, this cannot be expected as a regular policy. I think that we ought to be careful not to give by over-emphasis any excuse for misrepresentation of our claim for scientific freedom.[41]

Moreover, the definition of academic freedom did not extend to Communists.

> I myself have in the past been strongly critical of infractions of academic freedom on the grounds of belief or action having nothing to do with the victim's academic function . . . But the claim of Stalin's Communism to conduct its insidious and proselytising campaign under the shelter of an academic or official freedom . . . to concede it, I suggest, is simply to yield a position of advantage to the attack on 'freedom of science' which it is our business to defend.[42]

The new view of the scientific community depended, in fact, upon the rigorous exclusion of dissenting voices. By 1950 the SFS believed that this had happened. 'Official opinion', wrote Sir Henry Dale, 'does not seem likely now to be so susceptible . . . to insidious infiltration of Marxist ideas concerning science and its organisation'.[43] In fact the 1950s saw a proliferation of societies like the SFS — for example the Committee on Science and Freedom (1954). The conviction that Marxist infiltration in science had been pushed back was expressed at gatherings like the Milan Conference of that year. 'Marxism like National Socialism has lost its appeal to the Intellectuals.'[44]

The rhetoric of freedom in science was, in fact, accompanied by increasing restrictions upon it. Some of these in the case of the USA were of draconian proportions. Much to his surprise and chagrin Polanyi found himself caught up in these when in 1952 he was refused a visa to the United States.[45] This prevented him from taking up a chair in Social Philosophy at Chicago and, as he wrote, placed him for a time in a precarious financial position. Polanyi felt there were a number of possible reasons for his exclusion. He had attended student meetings in Hungary in 1908 though these were not, he claimed, political. In particular he had been a member — for a short time — of two organisations, the Society for Free German Culture during the Second World War and shortly after it, the Society for Cultural Relations with the USSR. In both these organisations he acted as propagandist against the views expressed in them. Finally Polanyi wondered whether the refusal might be the result of the malice of some Hungarian Communist in the United States with a long memory for, as Polyani protested, 'not even in my earliest youth have I had any left-wing leanings and since 1917 I have been an active fighter against Communism both in word and writing.'[46]

In spite of the reality of increasing state control, the image of the scientific community as independent, homogeneous and self directing became increasingly powerful in the late 1940s and early 1950s. Partly this reflected the ambition of many scientists for a high occupational status similar to that of other professions — a resistance to the view which saw them as another white-collar proletariat. Polanyi's interpretation of academic freedom as 'the right to choose one's own problem for investigation, to conduct research free from any outside control and to teach one's subject in the light of one's own opinion'[47] had professional as well as political overtones. It implied that the scientist like the doctor or lawyer had control over his own professional conditions and contracted out his services. In an article called 'The republic of science' written for *Minerva* in 1962, Polanyi clearly stated this. 'My title is intended to suggest,' he wrote, 'that the community of scientists is organised in a way which resembled certain features of a body politic and works according to economic principles similar to those by which the production of material goods is regulated.'[48]

Strangely enough in spite of the hostility expressed towards Marxist economic determinism, this view owed more to vulgar economics than Bernal's in *The social function of science*. What it presented however was a more acceptable version of the economy — one reminiscent of nineteenth-century political economy —

in which there is a natural equilibrium in society, in which social effects are simply the sum of individual action and in which hierarchy is the result of merit rewarded.

This view of the scientific community had another powerful persuasive appeal. It reintegrated science within the ideological parameters of Western society at a time when politicians' distrust of science and scientists was at its height. According to Polanyi not only was science not in contradiction to Western values, in fact it represented the strongest expression of them. The scientific community represented democratic consensus, 'deriving its coherence from its deep common rootedness in the same scholarly tradition'.[49] It combined this with individualism — the motor of progress and development — especially economic individualism. It was a smaller version of the market economy of the West. Polanyi wrote:

> What I have said here about the highest coordination of individual scientific efforts by a process of self coordination may recall the self coordination achieved by producers and consumers operating in a market. It was indeed with this in mind that I spoke of the 'invisible hand' guiding the coordination of independent initiatives to a maximum advancement of science, just as Adam Smith invoked the 'invisible hand' to describe the achievement of greatest joint material satisfaction when independent producers and consumers are guided by the prices of goods in a market.[50]

Polanyi's 'Republic of science' expressed the political values of the West as he saw them. As in the scientific community, the politics of the West are a combination of freedom rooted in tradition. This 'rejects the dream of a society in which all will labour for a common purpose determined by the will of the people, for, in the pursuit of excellence, it offers no part to the popular will and accepts instead a condition of society in which the public interest is known only fragmentarily and is left to be achieved as the 'outcome of individual initiatives aiming at fragmentary problems.'[51]

Even the problems of science were re-defined by this idealised picture. If science were the product of individuals who sought recognition for their intellectual efforts then a conflict could ensue between the consensus of the scientific community and the individualism of its members. In an earlier address Sir Henry Dale had referred to these problems.[52] Personality clashes and rivalry between scientists endangered the consensus. The existence of prestigious elites circumscribed equal opportunity. Secrecy blocked careers. Occasional publicity-seekers contravened the rules of free exchange or tried to pass off shoddy goods as the authentic currency of the market. If Polanyi had gone further in delineating the political economy of science he could have pointed to the existence of monopoly and imperfect competition distorting the scientific market.

This analysis displaced the views of Bernal. In the first place the state disappeared as a component necessary for the understanding of science. Bernal may be criticised for an over-optimistic view of the way in which the state and science would be integrated. But the existence and importance of the state was at the heart of his view of the direction of scientific development. The view that succeeded his simply wrote out the state at a time when its influence was increasing. Secondly Bernal had made science dependent on general social and economic conditions. This led him to a too easy acceptance of Lysenko's claims to have made biology socially useful and fitted to the revolutionary conditions of the USSR. But it also provided part of the basis upon which Lysenkoism could be explained. The alternative view simply disguised the social conditions which affected science making it into a self-regulating system which produced, by internal dynamics, its own structures. Incidentally, it was also forced to resort to notions of 'evil' and 'irrationality' to explain Lysenkoism. Finally, Bernal offered a view of the scientist which had more in common with their actual experience of working in large-scale enterprises than the rather elevated professional vanity contained in Polyani's description of the character of scientific work.

In fact the numbers of scientists were increasing in Britain in the 1950s. By 1959, there were approximately 25,334 in industry and 6,352 in government. But of the industrial scientists a considerable number were employed on defence work or other government projects subcontracted to private industry. There was, therefore, a great deal of state intervention, control over projects, and over the products of scientific research and also security measures to protect industrial and defence secrecy. This rather than a free market in ideas was the typical experience of the majority of scientists.

The Lysenko controversy acted as one component of the ideological counter-attack against the left. P.W. Bridgman writing in 1952 claimed that the work of the SFS 'has been so thoroughly accomplished that the Society is now rather inactive'.[53] Though satisfaction at this stage of affairs proved premature[54] none the less it was true that a different ideological consensus had emerged. This was strongly rooted in the notion that certain activities were not legitimate for scientists. It re-interpreted their work to make it acceptable to the traditions of Western political democracy.

Notes

1. H.J. Muller, 'The crushing of genetics in the USSR', *Bulletin of Atomic Scientists*, vol. 12 (1948), pp. 369-71, p. 369.
2. *Bulletin of Atomic Scientists*, vol. 5 (May 1949).,
3. John K. Jacobs, 'The scientist and military research', *Science and Society*, vol. 2 (1947), p. 75.
4. Margaret Gowing, *Independence and deterrence: Britain and atomic energy, 1945-52*, 2 vols., (Macmillan, London, 1974), vol. 1, pp. 171-2.
5. Ibid., vol. 1, pp. 115-16.
6. *Manchester Guardian* (5 November 1948), in a review of Blackett's book *The military and political consequences of atomic energy* (Turnstile Press, London, 1948)
7. Peierls papers, A1, Peierls to Walter Stanners (29 August 1946)
8. Ibid. F1-11 Peierls to Sir John Anderson, (29 May 1946).
9. Quoted a year later in *Atomic Scientists News*, vol. 2, no. 1, 1948, p. 98.
10. AScW memorandum, 'The international control of atomic energy' (August 1947), statement issued by the executive committee.
11. Peierls papers F1-11, Atomic Scientists Association, memorandum to council and vice presidents (28 November 1947)
12. Quoted in *Atomic Scientists News* vol. 2, no. 1 (1948), p. 33.
13. Ibid, p. 66.
14. Peierls papers, A1, on Nunn May and Fuchs (undated memorandum, possibly 1950).
15. Peierls Papers, F1-11, Kurti to Peierls (1 January 1948), newspaper cutting included.
16. Peierls Papers, A1, on Nunn May and Fuchs (undated memorandum, possibly 1950).
17. Peierls Papers, F1-11, Peierls to Mott (21 February 1950).
18. Peierls Papers F1-11, Skinner to Kurti (26 February 1948).
19. *Atomic Scientists News*, vol. 2, no. 2 (1948), p. 93.
20. Ibid.
21. Blackett Papers, J17, Burhop to Blackett (5 November 1948).
22. Peierls Papers, F1-11, Peierls to Cherwell (22 March 1950).
23. *Atomic Scientists News*, vol. 3, no. 5, May 1950, p. 108.
24. Cherwell Papers, D14, Cherwell to Peierls (15 February 1951).
25. Ibid, Peierls to Cherwell (19 February 1951).
26. *Atomic Scientists News*, vol. 4, no. 3 (January 1951), p. 79.
27. Ibid.
28. *Atomic Scientists News* vol. 1, no. 1 (September 1951), pp. 24-5; this

was the first issue of the new 'academic' series.

29. Peierls Papers, F1-11, Champion to Peierls (14 February 1950).
30. N.W. Pirie, 'Soviet genetics', *The Scientific Worker*, vol. 3, no. 6 (December 1948), pp. 28-9.
31. Angus Bateman, 'The Soviet scientific controversy', ibid, vol. 4, no. 3 (June 1949), p. 26.
32. G.T. Walker, London, ibid., vol. 4, no. 4 (August 1949), p. 29.
33. Douglas McClean, ibid., vol. 4, no. 1 (February 1949), p. 24.
34. J.W. Addie, Hon. Sec. North West Area, ibid., vol. 4, no. 2, (April 1949), p. 27.
35. *The Scientific worker*, vol. 4, no. 3 (June 1947), p. 21; the TUC circular on Communism in Trades Unions 'Defend democracy' had been circulated around the major unions including the AScW.
36. 'Report on the Lysenko controversy' (London, AScW 1951), p. 9. The Executive Committee set up an investigating committee in 1949. This was reported in the same issue of *The Scientific Worker*, in which the Executive denied Communist Party influence in the union.
37. Ibid., p. 8.
38. Ake Gustafsson, 'Marxist genetics at the Stockholm Botanical Congress 1950', *Journal of Heredity*, vol. 42 (1951), p. 69.
39. Margaret Gowing, *Independence and Deterrence*, vol. 1 (1974), pp. 48-57.
40. Earl of Birkenhead, *The prof in two worlds* (Collins, London, 1961).
41. Sir Henry Dale, SFS Occasional Pamphlet, no. 11 (March 1951), p. 7.
42. Ibid.
43. Ibid., p. 8.,
44. Report on the Milan Conference, Committee on Science and Freedom (Congress for Cultural Freedom, Paris, 1954) p. 7, and 8. The committee was established as an international organisation in 1954 to carry on discussions begun in 1953 at the Hamburg Conference. It was an international version of the SFS. Later on it becomes connected with the foundation of the periodical *Minerva* in 1962.
45. *Bulletin of Atomic Scientists* vol. 8, no. 7 (October 1952).
46. Ibid., p. 223.
47. Michael Polanyi, 'The foundations of academic freedom', SFS Occasional Pamphlet no. 6 (1947), p. 4.
48. Polanyi, 'The republic of science', *Minerva*, vol. 1, (1962) no. 1, pp. 54-73
49. Polanyi, 'The foundations of academic freedom', p. 13.
50. Polanyi, 'Republic of science', p. 56.
51. Ibid., pp. 72-3.,
52. Sir Henry Dale, SFS, Occasional Pamphlet no. 11 (March 1951), p. 7, fn 84.
53. *Bulletin of Atomic Scientists*, vol. 8, no. 7 (October 1952), p. 229.
54. The AScW kept going and, in regard to atomic weapons policy, felt they had achieved some success when the Trades Union Congress voted for unilateral disarmament. According to its journal, 'To those of us in the AScW who can remember when our association stood alone in its

opposition to the manufacture and testing of atomic and nuclear weapons, this was indeed a momentous change' *The Scientific Worker*, vol. 6, no. 6 (November 1960).

4
Race: a New Beginning

In 1948 *American Anthropologist*, journal of the American Anthropological Association, published a letter from a colleague in Germany, Dr Franz Termer, director of the museum für Völkerkünde in Hamburg. Termer's letter records both the considerable physical destruction experienced by German museums and collections due to Allied bombing and the personal disruption experienced by German anthropologists under the occupation.

> As far as ethnology in the rest of Germany is concerned, I know, briefly, the following: the Berlin Museum has been destroyed. Krickberg became director, Nevermann custodian, Gelpka (Asia) shot himself. Baumann was discharged and is living miserably near Berlin, Disselhoff was captured by the English. He was freed, but his reappointment is impossible. Leipsic. The Museum is half destroyed with great losses, especially in the African, European and American collections. The new director is Erkes, a Sinologist. Krause had already been discharged by the Nazis in 1944. All the others, like German Damm, were also discharged.[1]

However, the blows to the prestige of German science in the period 1945–6 were even greater and more profound than physical destruction and military defeat. From 1945 to the end of 1946 the allies reached German-occupied territory from both East and West. Concentration camps were liberated and details of the situation in them reached the newspaper reading public in Britain and the USA. Court martials were held in British-occupied territory at Belsen and Auschwitz in September 1945.

Details about the use of human subjects in experiments were

made known. On 26 September 1945, experiments on artificial insemination at Auschwitz were reported. On 12 July 1946, experiments on human subjects for a preventative vaccine for typhus at Buchenwald were described in British newspapers. On 4th October 1946 accusations were made of experiments on Greek girls to find an efficient method of sterilisation and on 5th October, further accusations emerged of gassed bodies being sent to a Professor Hirt at Strasbourg University for experimental purposes. On 13th October, the notorious twin experiments carried out by Mengele at Auschwitz were reported, and on the 14th October the malaria experiments conducted by Karl Klaus Schilling at Dachau. A Major Rasche, a Luftwaffe doctor, was accused of conducting endurance experiments on prisoners. On 9th December 1946, the trial of 15 Nazi doctors opened and on 29th August 1947 they were sentenced.[2]

Since 1942 it had been known that experiments on the effect of phosphorus burns had been conducted on Russian prisoners of war and, at the Nüremburg Trials which lasted from 1945 to 1946, it was reported that the German Admiral Canaris had made an official protest about this contravention of the Geneva Convention on prisoners of war. Knowledge of the full extent of these experiments had to await Germany's defeat. The persistent abuse of prisoners of all nationalities by medical men and scientists seemed, even amid the terrible conditions of deprivation in concentration camps, to be a particularly calculating and cold form of cruelty. Overshadowing all this was of course, the revelation of the systematic killing of certain groups and nationalities and the particular suffering of the Jewish people. Jews had been subjected to all the varying cruelties of experimentation, physical abuse and deprivation. Eventually it became clear that they had been subjected to a programme of systematic extermination. The question for many was why Germany, a nation renowned for its culture and science, had reached this point of degradation.

In the 1930s, an extensive network of Eugenics Societies existed in most of the major countries of the world. The Eugenics Education Society of London was founded in 1907. This society eventually incorporated many British provincial societies which sprang up about the same time. Equivalent eugenic associations followed in Germany, the United States and France. In the inter-war years eugenic societies under various names could be found in most

Scandinavian countries and as far apart as South America, Japan and China.[3]

All of these societies were pledged to improve the human stock of which their respective nations were composed. Certain forms of reproductive technology, such as birth control and sterilisation, were available in the inter-war years to aid the eugenic project. These supplemented other methods which, in the early years of the twentieth century, had been widely discussed as a means to control the fertility of the mentally deficient and other groups perceived as 'social problems'. These included institutionalisation, immigration control, financial incentives or disincentives for fertility. In some cases, the penalty of death or euthanasia was seriously discussed as a remedy for hereditary unfitness but most eugenicists hesitated at advocating this.

Every country revealed the determination of its own particular social and political structure in its eugenics movement. In France a relatively declining population led its eugenicists to a strong commitment to pro-natalist policies. In Britain class was the chief preoccupation and the majority of eugenicists were pre-occupied with redressing the differential birth rate between the middle and working classes.[4]

In the 1920s and 1930s sterilisation became a major issue for the international eugenics movement. Sterilisation laws were adopted in many American states in the 1920s and 1930s and in some Canadian provinces — in Alberta in 1928 and British Columbia in 1933. They were also passed in Tasmania and New Zealand. A sterilisation law was passed in Sweden in 1934 (in force from 1935). Estonia adopted one in 1937, the Mexican state of Vera Cruz in 1932. It was the official policy of both the British and Czechoslovakian societies in the 1930s although both were unsuccessful in their attempts to get a sterilisation law passed. In 1933 after the advent to power of Hitler, a sterilisation law was passed, which was enforced in 1934.

It was impossible to advocate such a programme of eugenics without having a set or scale of social and political values which determined who were the 'fit' and who the 'unfit'. Moreover it was also necessary to believe that social progress could be achieved through the manipulation of heredity. In the 1920s, a wide variety of political beliefs and values could, potentially at least, attach themselves to a eugenic credo. In *Man and superman*, George Bernard Shaw envisaged a utopia of the hereditarily superior who would realise the socialist dream of equality and community. Others believed that, in the hands of women, the control of reproductive power would be a powerful weapon to redress the inequality between the sexes. Some believed that the scientific manipulation of sex which eugenics promised would dethrone Victorian patriarchy and, therefore, allow the free development of experimental and bohemian life styles.

By the late 1920s these visions of eugenics as liberation became increasingly difficult to realise. Apart from mavericks like the American geneticist H.J. Muller who believed that true hereditary worth would only manifest itself in conditions of social and economic egalitarianism and left for the Soviet Union in the 1930s to prove his point,[5] the majority of eugenicists were on the right. Although intellectuals, and 'progressive' intellectuals in particular, were convinced of their own importance — as measured by their published books, lectures and articles — social reality in the form of the economic depression of 1929-31 and the subsequent political struggles, proved more important in influencing the direction eugenics took. Thus, in nearly all the countries in which eugenics movements existed, the eugenic movement was harnessed to reinforce certain values; the importance of nationalism and national strength; the racial superiority of some groups over others; the hereditary unfitness of the poor; the uselessness of ameliorating social conditions through economic reform; the dangers of socialism; the threat posed by alien groups to culture and national strength.

Eugenics societies were never exclusively composed of scientists or medical men. Their membership included a variety of people — politicians, businessmen, social workers and the usual sprinkling of busybodies found in organisations of this type. Nor should the eugenics movement be identified exclusively with the institutions set up to propagate eugenics. The gospel of eugenics spread into other quasi medical, charitable or social work institutions. It had friends and supporters in the legislatures and among the politicians of the countries in which it existed; it even had 'popular' appeal, as evidenced by the kind of favourable coverage which, in Britain, for example certain newspapers such as the *Morning Post* and *Daily Mail* gave it. Nonetheless eugenics relied on 'scientific' or medical justifications for its existence and on scientists and doctors for an authoritative interpretation of its tenets. It was therefore a nodal point at which political and social belief and the practice of science met.

This can be most clearly seen in a classic textbook of genetics and heredity, written in the 1920s by three German biologists,

Eugen Fischer, Erwin Baur and F. Lenz. This textbook *Human heredity* ran into three editions the third of which was translated into English in 1931.[6] This book contained a classic exposition of Mendelian genetics, which was used by students for the succinct accounts it gave of the 'new' science. At the same time the authors clearly had a social mission. This was to preach the doctrine of hereditarianism: 'If a nation or a particular stratum of a nation has an inferior hereditary equipment (the Negroes for instance in the United States of America), education and cultural influences may improve members of the nation or the stratum but they cannot alter the stock.'[7] This, the authors wrote, was an opinion which 'conflicts sharply with pre-conceived notions which are widely current. But neither the biologist nor the physician must allow his scientific insight to be clouded by such political prejudices'[8] (presumably the idea that environmental reform can lead to social improvement).

Baur, Lenz and Fischer also believed in the importance of eugenics and of race. Much of *Human heredity* was concerned with the reconciliation of traditional physical anthropology with Mendelian genetics. In addition this discussion was interspersed with comments about the relationship between national and racial strength:

> We are coming to recognise, more and more clearly that racial factors and especially hereditary mental endowments, although they work (it cannot be too often repeated) in conjunction with other factors, are among the most influential in determining the course of a nation's history . . . What historians regard as degeneration, sickness and ageing of a nation, what they look on as the decline of a nation are the outcome of a reversed selection of the racial constituents of the people concerned.[9]

Baur, Lenz and Fischer also shared British eugenicists' concern with the proliferation of the less well-endowed at the expense of the better-endowed among populations of similar racial stock.

Anglo-American re-interpretation of eugenics have often foundered, as in the case of Baur, Lenz and Fischer, on two misconceptions. First these authors did not confuse race with nation and were well aware that nations were political and cultural creations. But this does not exonerate them from the charge of racism. They clearly believed in the supremacy of the Nordic races and that a nation in which the Nordic race loses its predominance and is threatened by 'alien' elements would go into decline. Second, they were perfectly prepared to grant valuable qualities to non-Nordic races. Jews, for example, they considered to have a capacity for intellectual abstraction and therefore to succeed in philosophy and mathematics. But again whilst each race 'scored points', the Nordic was assumed to be superior. Moreover — and this is where certain trends of European racial thought clearly distinguished itself from Anglo-American beliefs — racial superiority was not exclusively mental or intellectual. Nor was it necessarily rewarded by material or social success. They believed the Nordic was emotionally and spiritually superior. Their view of him recalls Nietzsche's superman, whose exceptional character often led to spiritual and social isolation. Baur, Lenz and Fischer depicted the Nordic as aloof, full of aristocratic reserve and hauteur, with a poor capacity for co-operation, for the toleration of monotony and routine and with little human sympathy. These qualties which, in an American scale of values, might be considered faults were, to the authors of the textbook *Human heredity* evidences of moral and intellectual superiority.

After Hitler's accession to power in 1933, the German eugenics movement became more racial and anti-semitic. This worried some German eugenicists and it also alienated eugenicists in other countries. However, this did not happen to the extent that the German eugenics movement found itself isolated. On the contrary much of what passed for eugenics in Germany in the 1930s was recognisable to eugenicists in other countries and was, mostly, warmly welcomed.

In 1937 Otmar von Verschauer, a proponent of Nazi eugenics who headed the Frankfurt Institute of Hereditary Biology and Racial Hygiene, set out four aims for German eugenicists. They wanted to prevent the race being destroyed, and to do this their objectives included: (1) Avoiding restricted selection (to counteract which von Verschauer cited the pro-natalist policies established in Germany to prevent population decline; i.e. the encouragment of marriage and motherhood); (2) Avoiding unfavourable selection — that is, the breeding of the 'unfit' — which would be prevented by the spread of birth control, sterilisation and euthanasia; (3) Avoiding the injury done to the race by the passing on of bad physical or mental traits (genetic counselling and some of the measures outlined above would deal with this); and (4) Preventing the introduction of foreign elements, which referred to the regime's anti-semitic policies.[10] Only the last of these points raised eyebrows among eugenicists in Britain and America. But even so, not everyone was prepared to dismiss Nazi eugenics on this account. Paul Popenoe of the Human Betterment Foundation of the USA gave the German population policies a sympathetic hearing in 1934. (By then a very wide-ranging policy of compulsory sterilisation of the mentally deficient had been adopted in Germany)

> But Hitler himself though a bachelor has long been a convinced advocate of race betterment through eugenic measures. Probably his earlier thinking was coloured by Nietzsche but he studied the subject more thoroughly during his years in prison, following the abortive revolutionary movement of 1923. Here it is said he came into possession of the two volume text on heredity and eugenics by E. Baur, E. Fischer and F. Lenz which is the best known statement of eugenics in the German language and evidently studied it to good purposes.[11]

Even C.P. Blacker of the Eugenics Society of London wrote to a correspondent in February 1939 that, whilst understandably disturbed by aspects of German policy, 'I think it would be a good thing if the impression were removed that as a committee we disparage the results of the German policy. For my part I regard these as substantial and indeed remarkable'.[12]

Most British and American eugenicists and many in other European countries distanced themselves from anti-semitism. The problem was that Nazi practice reflected on other areas of 'respectable' eugenic practice. In 1947, Lord Moran sent C.P. Blacker a file on Nazi medical experiments because of their apparent eugenic content. In particular he asked him to comment on experiments to find a quick and effective means of mass sterilisation. Sterilisation had preoccupied the eugenics movements in other countries. In some, voluntary sterilisation measures had been enacted and much of the energy of the British eugenic movement went into trying to persuade Parliament and public opinion of the importance of sterilisation, in particular as a defence against the 'ever rising tide of feeblemindedness'. The eugenics movement had a particular animus towards the mentally deficient and in Germany this had been carried as far as it could. *The Times* (17th January 1946) reported on the projected trial at Nüremburg of Wilhelm Frick who, as Minister of the Interior, played a leading part in the orders to do away with mentally deficient people and other 'unworkables'. The fact that these were not prisoners of war nor Jews but individuals taken from ordinary German homes had caused comment among Germans at the time. The Bishop of Limburg, Dr Hils Frich, had protested against the policy in 1941, along with other religious leaders. 'The facts', the Bishop wrote, 'had become notorious.'

> Several times a week nine buses arrived in Hadamar with a large number of victims. School children knew the vehicle and would say 'There comes the murder box again' and the citizens of Hadamar would watch the smoke rise from the chimneys with the ever present thought of the miserable victims. Children calling each other names would say, 'You're crazy, you will be sent off to the baking ovens in Hadamar' and the old folks said, 'Don't send me to a state hospital. After the feebleminded have been finished off the next useless eaters will be the old people'.

The plan, so the report said, had been formed in 1940 by Dr Wilhelm Frick and Dr Conti, chief physician of the Reich, who committed suicide at Nüremburg. It was alleged that 200,000 mentally deficient people and 75,000 of the aged had been done away with.[13]

The twin studies conducted by Mengele at Auschwitz under the general academic direction of Otmar von Verschauer had relevance to eugenics. Twins had acquired a particularly significant place in eugenics as a test to determine whether an hereditarian explanation of mental ability and character was correct. If identical twins had the same IQ, or criminal propensities (even when reared in different environments) then the argument for hereditary determination of character and ability was strongly supported. Again, the Nazi had ignored all countervailing moral principles in their conduct of these twin experiments. Even so, it was not just facist anti-semitism on trial in the 1940s but also the logic of eugenics itself — in particular, the practice of eugenics in singling out certain social groups as obstacles to national strength or social progress, focusing dislike, even hatred, on them and removing them from the protection of the moral law as it applied to the 'fit' and healthy.

Professor Zollschan had failed in his attempt to secure international co-operation in the fight against racism in the 1930s.

However, in 1946 the United Nations Educational, Scientific and Cultural Organisation (UNESCO) came into being. Julian Huxley was its first director, followed by Jaime Torres Boder. One of its objectives was the regeneration of education in devastated Europe and also in other areas of the world. With this in mind it conducted surveys to estimate the educational needs of countries, it organised collections of book and other educational material for schools, provided scholarship and grants. But UNESCO had other objectives. Taking its cue from the United Nations Charter of Human Rights, finally formulated in 1948, it proclaimed its objective to be re-education for democracy and for equal treatment of peoples. Its director of the rconstruction and rehabilitation programme, Dr Drzewieski, a Pole, clearly felt that some at least of the troubles of the last decade could be put at the door of the educators, a kind of *trahison des clercs*.

> The forces of evil in many countries exploited the national neutrality of educators and cultural leaders. They appealed to the best feelings of the adolescents and youngsters who were looking for simple ideas and realistic activities. Then the educators began to develop in the youngsters feelings of nationalistic conceit and contempt for other races treated by them as inferior. Hence (the) victory of Fascism and naziism.[14]

UNESCO therefore decided to go on the offensive. It pledged itself to a through investigation of race and racial ideology to dispell false ideas and to counteract nationalism and fascism. The plan laid down by the Organisation proceeded from a resolution (116 (VI) B (iii)) adopted by the United Nations Economic and Social Council at its Sixth Session asking UNESCO 'to consider the desirability of initiating and recommending the general adoption of a programme of disseminating scientific facts designed to remove what is generally known as racial prejudice.' In reply the fourth session of the UNESCO general conference adopted three resolutions — to study the question of race and collect scientific material on the subject, to disseminate it, to prepare an educational campaign in connection with it.[15] The Department of Social Sciences under Dr A. Ramos convened anthropologists, psychologists and sociologists to do this at a meeting from 12th to 14th December 1949.

The UNESCO statement was issued on 18th July 1950. It was drafted by Ernest Beaglehole (New Zealand), Juan Comas (Mexico), L.A. Costa Pinto (Brazil), Franklin Frazier (USA), Morris Ginsberg (Great Britain), Dr Humayin Kabir (India), Claude Levi Strauss (France) and Ashley Montagu (USA). But revisions were made after criticism by Professor Hadley Cantril,

E.G. Conklin, Gunnar Dahlberg, T. Dohzhansky, L.C. Dunn, Donald Hager, J.S. Huxley, Otto Klineberg, Wilbert Moore, H.J. Muller, Gunnar Myrdal, J. Needham and Curt Stern. Whereas the first drafting committee was largely, though not exclusively, composed of cultural anthropologists and sociologists, the second advisory committee which amended it contained a high proportion of geneticists.

On the whole, these committees represent the centre or left of centre, the liberal wing of middle-class society with a sprinkling of the far left. They represented, in fact, the political consensus historians have called the ideology of the 'resistance' which emerged out the defeat of Hitler's Germany in 1945. The document produced by UNESCO reflected this. It attacked the use of the term 'race' to describe various groups or nationalities. It argued that the term 'race' should properly speaking be applied only to breeding populations in which some genetic variation has arisen by virtue of geographical or cultural isolation. It argued that the variations which give rise to distinctive physical differences between peoples are a small proportion of the total number of genes. Most genes we have are shared in common with other groups. It also argued that such racial groups are 'dynamic' — their genetic composition changes over time and by the process of intermingling with other genetic groups.

As well as covering genetics the statement went on to examine mental classification, i.e. IQ tests. It said, 'It is now generally recognised that intelligence tests do not in themselves enable us to differentiate safely between what is due to innate capacity and what is the result of environmental opportunities.'[16] It stated that there was no relationship between culture and genes nor are traits or personalities genetically determined. It defended 'hybridisation' or race crossing. In the opinion of the committee there was 'no biological justification for the prohibition of inter-marriage'.[17] It quoted Darwin on the importance of social sympathy in human evolutionary development.

This statement was aimed at a world free from the kinds of prejudices which had marred the 1930s. Those who drafted it hoped their statement would help one emerge. But it ran into immediate difficulties. A good proportion of anthropologists were not prepared to accept it for a variety of reasons. Shots over the bow of the UNESCO statement were discharged in the journal *Man*, organ of the Royal Anthropological Institute of Great Britain. An advanced copy of the UNESCO document had been obtained and circulated to Professor le Gros Clark, H.J. Fleure, Dr Harris, Dr Osman Hill, Sir Arthur Keith, D. Morant, Myres Tildesley, Dr J.C. Trevor and Professor S. Zuckerman. According to *Man*, 'Most of these replied at some length and their comments made it perfectly clear that certain passages in the statement were far from commanding universal agreement.'[18] So much was this the case that a letter was despatched to *The Times* of London on 15th August 1950. In response to this Metraux, who was head of the division for the study of race problems at UNESCO wrote to *Man* telling them of his decision to convene a further conference of experts. An editorial in May 1951 in *Man* argued that whilst the subjects discussed at this could include polygenism (i.e. the idea that the races of man are separate species) neither Michurinism (Lysenkoism) nor any contribution from the Soviet bloc was acceptable.

This editorial signals a far more aggressive attitude on the part of some anthropologists and geneticists to the attempts at the 'reconstruction' of their discipline by UNESCO's humanists. It was now the height of the cold war and because of this some were prepared to attack the legacy of 'resistance', to denigrate egalitarianism, and to re-inject a hereditarian pessimism into discussions of race. When the conference convened on 2nd to 9th June 1951, this became evident. As well as the participants, UNESCO had solicited statements and comments from 92 scientists concerned with genetics and anthropology, of whom 73 replied — some at great length. Of these replies, 23 were recorded as being unreservedly in favour. Some like the British geneticist L.S. Penrose, believed the statement did not go far enogh. He thought it was now possible to drop the term 'race' altogether. However, 50 ventured criticisms which ranged from being mild to outrightly hostile.[19]

The hostility expressed by most geneticists was to the sweeping assertions made by the statement rather than to its spirit or intentions. For example H.J. Muller, the American geneticist, believed that the statement represented the views of a minority of practising geneticists. This was because the majority, including himself, were still convinced that mental faculties had a significant genetic component. Muller considered that this assumption was built into the research programme of most geneticists. This did not mean, he argued, that the spirit of the UNESCO document could not be preserved. 'It would be a tragic mistake to suppose that the above realistic scientific viewpoint leads to the conclusion that race prejudices are justified . . . The essential points are that the different racial groups a) are enough alike genetically b) are capable of being so much influenced in mental development by cultural and other environmental factors and c) contain such important individual genetic differences for psychological traits within each one of them that all of them are capable of participating in and cooperating fruitfully in modern civilisation (as has also been empirically demonstrated).'[20]

However a hard core of geneticists and anthropologists saw the UNESCO document as a provocation and they said so. Prominent among these were those British geneticists and anthropologists with a history of involvement in eugenics. British eugenics had been strongly focused on class but this was determined by the political situation in Britain. When forced to concentrate on wider issues of racial differences, the leading lights of British eugenics proved as convinced that a racial hierarchy existed as they had of one based on class or occupational differences.

C.D. Darlington was joined by R.A. Fisher, a famous geneticist and former secretrary of the Eugenics Society, in criticising the UNESCO document. Darlington considered that human groups, 'differ in their innate capacity for intellectual and emotional development'. He insisted that the chief political problem facing the world was learning to share the resources of this planet amicably and that 'this problem is being obscured by entirely well intentioned efforts to minimise the real differences that exist'. The idea that these racial differences were distinct but of equal value, he argued, was a fiction.[21] Similarly C.P. Blacker, secretary of the Eugenics Society, attacked the UNESCO statement in the *Eugenics Review*. He considered that 'It is at least possible in view of the acknowledged existence of inborn and physical differences between races, that subtle but nonetheless important inborn differences in cognition, affection and conation, as yet unamenable to psychological tests may exist; or that differences common to all humanity may have varying distributions among difference races.'[22]

J.R. Baker was even more adamant. He was a life-long eugenicist and a friend of C.P. Blacker. He attacked the UNESCO document in a Society for Freedom in Science pamphlet saying, 'the definite statement is made (on p. 65) that the mental and cultural differences (between the various groups of mankind) are dependent on environment, not on racial traits. Now some scientists who are qualified to hold an opinion on this subject think it is probable that there are innate mental differences between the races of mankind, they do not consider that there is any satisfactory proof that the Australian aborigines (for instance) have, on the average, exactly the same inborn intellectual potentialities as the peoples of Europe. It is not legitimate for a government body like UNESCO to say otherwise.'[23]

The British eugenicists received support from other more surprising quarters. Although the events of the war had disrupted German anthropology and genetics, gradually some who had been prominent in pre-war German genetics crept back into positions of influence. The process of denazification was cursory and in an area so relatively esoteric to the occupying forces in Germany as the biological and human physical sciences, careful vetting was unlikely.[24] Observers of the German academic scene at the time were critical of the process of reconstruction within German academia, which they believed to be incomplete and haphazard. To some extent their fears were confirmed by the reappearance of Lenz and Fischer. Although heavily compromised by their association with pre-war German eugenics, both were invited to submit comments on the UNESCO document. They were highly critical of it, with one crucial exception. Lenz carefully made a clear denunciation of anti-semitism. This he regarded as a mistake and an aberration. However, he criticised the UNESCO statement because it 'disregards not only the enormous hereditary differences between men but also the absence of selection as the decisive cause of the decline of civilisation and it therefore runs counter to the science of eugenics'.[25]

Some Germans, Professor W. Scheidt for example, asserted that the German contribution would be discounted because of the past, 'Any objection which Germans might raise to this statement (which is in flat contradiction with the present policy of nearly all UNESCO's member states) would probably be misconstrued as a survival of Nazi ideas.'[26] In fact considerable latitude was given to the German contribution. Fischer, for example, compared the UNESCO document to 'the National Socialists' notorious attempts to establish certain doctrines as the only correct conclusions to be drawn from research on race, and their suppression of any contrary opinion.' He also claimed the UNESCO document recalled 'the Soviet Government's similar claim on behalf of Lysenko's theory of heredity and its condemnation of Mendel's teaching. The present statement likewise puts forward certain scientific doctrines as the only correct ones and quite obviously expects them to receive general endorsement as such.'[27]

But another German geneticist, Hans Nachtsheim, pointed out

the UNESCO document when it was subsequently challenged by J.R. Baker at the Hamburg conference on science and freedom organised by the Congress for Cultural Freedom, the two cases were hardly the same. The UNESCO document was thoroughly discussed before it appeared, modified in the light of criticism and freely criticised without penalty by its opponents. He contrasted this with the situation in Germany in 1935. In that year, he, Nachtsheim, had defended Karl Felix Saller for an unorthodox theory on race in an article for a newspaper, 'But the newspapers of 1935 had not the courage to publish such an article. I therefore wrote a long leaflet to the National Socialist Minister, Frick . . . I remember a telephone call a few days later from the Ministry, to the effect that if I valued my personal freedom, I had better keep quiet.'[28]

It was not that certain geneticists and anthropologists resisted 'reconstruction'. In one clear direction they accepted it. Anti-semitism had become disreputable. Other races, however, remained vulnerable to prejudice. Hans Weinert complained about UNESCO's defence of hybridisation, 'In defence of prohibiting marriage between persons of different races I should like to ask which of the gentlemen who signed a statement would be prepared to marry his daughter for example, to an Australian aborigine . . . If it is true that all races have the same innate capacity for intellectual development then why is it that so far only the members of the white race have built up any scientific knowledge'.[29]

Some historians have interpreted the controversies over race in the late 1940s and early 1950s as being, in part, due to an increasing dislocation between genetics and the practice of anthropology.[30] Whereas physical anthropology traditionally defined race in terms of physical (and sometimes mental characteristics), genetics suggested a different approach. It implied that races were conglomerations of genes or gene pools. Within each gene pool, certain distinctive traits had developed. This was because of geographical or social isolation, genetic drift or selective pressures within the particular environment in which the gene pool was placed. But this also meant that most genes were shared by all races (races as defined by traditional physical anthropology). It also meant that variation in genes which might give rise to sometimes large differences in physical appearance could actually be quite small. It suggested that, in contrast, significant genetic differences might not be expressed phenotypically — i.e. might not manifest themselves in observable physical difference. Further most of the genetic variation found clustered in one gene pool could also be found, though much less pronounced, in other gene pools. In addition since all genetic heredity was subject to selective pressures, our genetic constitution was potentially variable, although there was disagreement about whether 'natural' selection was still operating or had ceased to be a significant influence over heredity. Faced with this, so the argument goes, the categories made by physical anthropologists were under attack and so too were the political assumptions about racial inequality which had been based upon those old categorisations.

It was certainly true that the arrival of genetics caused some heart-searching among physical anthropologists. One commented in 1951, 'With increasing interest in human genetics more and more anthropologists are succumbing to the criticism from geneticists that race should be based on genotype alone. Accordingly some are beginning to define race for the public in genetic terms coupled with statements to the effect that the former definitions are the product of an old-fashioned science employing misconceptions.'[31] An extreme example of this was of course, demonstrated by L.S. Penrose's belief, outlined in his submission to UNESCO, that the term 'race' should be abandoned altogether as misleading and inaccurate. As the 1950s progressed, awareness of genetics increased among physical anthropologists. A survey of the knowledge required by students, published in the *American Journal of Physical Anthropology* in 1971, placed genetics at the top of the list. 'Genetics', the article stated, 'came into prominence in the late 1940s and early 1950s.'[32] In addition, it is clear from the UNESCO document that developments in genetics were being used to 'liberalise' racial attitudes.

A corollary of the view that those who rule do so because they are wise is that prejudice melts in the face of advancing knowledge. Unfortunately, this is not true. There are some things more tenacious and deeply rooted than the pursuit of pure knowledge. One is the belief in social or racial superiority. The dismay felt by some physical anthropologists at the disruption caused in their discipline by the advent of genetics, was used at the time as evidence of the retreat of the racial beliefs of the 1930s. Moreover it was true that some anthropologists had based their belief in racial hierarchy on scientific ideas which were subsequently discredited. But the situation was more complex.

First, some genetics had begun to creep into physical anthropology prior to the 1940s.[33] Judging by the journals of physical anthropology, interest manifested itself in the 1930s in the link between the genetics of blood and racial groupings. It was observed that racial groups differed in the percentage of different blood groups they had. But this preponderance of blood groups in different races was only relative. Again similar blood groups could be found in all races. The field of genetic pre-disposition to disease and its geographical and racial distribution was also one which opened up in the 1930s. Thus whilst traditional racial categorisation still predominated, many anthropologists were aware that the naked eye and a pair of calipers were no longer as central in physical anthropology as they had once been.

Some of those most interested in the possibility of genetics in anthropology were also strong believers in race as a determinant of history. The British scientist, Reginald Ruggles Gates, for example, was a contributor to the application of genetics to anthropology in the field of blood groups. Some of his ideas, the belief that races were separate species, were idiosyncratic and became increasingly so in the 1940s. But Ruggles Gates clearly believed that genetics would not 'dish' race. Similarly Baur, Lenz and Fischer attempted in the 1920s to incorporate genetics into their racial classifications. They believed that the then 'new' genetics supported their views on race for three reasons. Firstly, because genes do not blend, they felt that racial hetrogeneity was ruled out. Racial features would always reappear in succeeding generations and be potentially separable in the human genetic constitution. Secondly, they argued that there must be some basis in genetic heredity for the anatomical differences noted by traditional anthropology, though this did not necessarily mean that one gene corresponded to one trait. They happily added evidence of genetic blood groups to the traditional racial categories. Lastly, they felt that, given the importance of genetics, there must be some genetic basis for mental differences as well as physical. Thus those who felt that the demon of racial thinking had been exorcised by genetics in the 1940s and 1950s were wrong.

Already in the 1930s, a group of anthropologists attacked racial thinking in their discipline. The UNESCO document of 1950 was, in effect, their testament. That they had the opportunity to state it was a product of the politics of the Second World War. Rightwing anthropologists tended to identify this group as a 'clique' whose mentor was Franz Boas and which included M. Herskovits, Ashley Montagu and Otto Klineberg. It was significant that the latter were American. In the United States, ethnic conflict and division was an important issue. In Britain, there were anthropologists and geneticists who attacked racial thinking, such as Julian Huxley and A.C. Haddon. They published a book on this subject, '*We Europeans*', in 1935. But Haldane, Hogben and others who attacked the right in the 1930s tended to be fixated on the question of class. Though they provided defences of egalitarianism and attacks on existing inegalitarian thought, they did not produce a body of thought so resolutely focused on race as the Americans.

Boas, a German emigrant to the United States in the 1890s, had entered the fray with attacks on the idea of a close correlation between anthropometric measurement and race. In his famous study of the German Jewish and German school population in the Germany of the 1880s, he revealed a considerable overlap and merging of physical characteristics between the two populations. He repeated these anthropometric experiments in later studies of 'primitive' races and reached similar conclusions. This put Boas at the head of an unofficial school of anthropology which, whilst not ignoring racial differences, tended to devalue them especially as explanations of cultural behaviour. This view reached its fullest expressions in Ashley Montagu's articles in the 1940s defending the role of cultural anthropology.[34]

Whereas Boas worked with the relationship between race and culture, Otto Klineberg concentrated on IQ. Most proponents of IQ have, rightly, been described as strongly in favour of the belief in the value of IQ tests in identifying social groups and races. But in the inter-war period, there existed a minority, of whom Hogben and J.L. Gray in Britain were two, who resisted this belief and either turned the conclusions on their head or attacked the validity of the tests themselves. Klineberg in 1931 tested selected racial groups in Europe (based on traditional classification) for performance in IQ. He did this to compare them with the performance of the groups tested by Davenport in the US Army in the First World War and to see whether, as in Davenport's tests, the Nordics performed better. Klineberg came to the conclusion that the most significant difference was the urban/rural divide. In the 1930s he went on to test the significance of this finding for the test scores of American blacks and again found urban acculturation a major factor in IQ performance.[35]

Those who wished to reformulate anthropology freed of its residual racism faced two problems. One was 'realism' or 'common-sense'. Our social relationships are set for us and, aware as we are of this, we do not treat a bank manager in the same way as a road sweeper. If we do, we are often punished for it by the indirect mechanism by which society teaches us who is important and who is not. Therefore as the outburst from some contributors to the UNESCO discussions shows, it is extremely painful and often appears absurd to behave as though social roles could be changed or were illegitimate or lacked moral authority. As one critic of genetics in physical anthropology put it, 'The public is not so easily fooled.'[36] Secondly, the critics of racism in the 1930s were politically up against powerful social forces, given the political complexion of Europe. They found it hard to escape acquiring the status of a minority and often, in the eyes of others, an eccentric minority. Thus Professor Zollschan tramped around Europe failing to launch his project of a concerted anti-racist stand among intellectuals. However, history sometimes catches us out. The events of the 1940s changed a great deal. For a while at least this gave the opening to those prepared to attack established beliefs and allowed the production of a document like UNESCO's statement on race.

The political change was felt everywhere, even if temporarily. Cyril Burt, for example, the British proponent of hereditarianism, was led in 1941 to a re-examination of his beliefs. He wrote to his sister about the German advance in the Soviet Union.

> The interesting thing is that under Stalin's Communistic organisation the Russians have done so well both in regard to manufacture, organising the Army and keeping the morale of the population so strong that they are willing to destroy their own homes and crops at Stalin's behest. After the fight with Finland, I thought that the Russian revolution had proved its failure as an organising power. As you know, my idea is that the labouring classes have less intelligence than the aristocrats and the businessmen. Consequently labour leaders (as you may hear on the English wireless) are evidently not a patch on the aristocratic families and on the businessmen when it comes to organisation. I thought, therefore, that Stalin had probably shot most of the intelligent Russians and that the rest had left, and that the country was being run by relatively unintelligent members of the Communist party.[37]

Burt was, however, in 1941 prepared to reconsider these points. But though his conversion was short lived and Burt retained his beliefs in the lower intelligence of the labouring classes, the relative egalitarianism of the 1940s in Britain — emphasising the equal contribution of all to the war effort, criticising the 1930s, promising more generous welfare and bringing in full employment — increasingly isolated him in the 1940s and 1950s.

Notes

1. *American Anthropologist* NS, vol. 48 (1946), p. 694.
2. *The Times*, 26 September 1945; 12 July, 4, 5 13 and 11 October 1946.
3. There is a growing literature on eugenics. For Britain and the USA, see Daniel J. Kevles, *In the name of eugenics, genetics and the uses of human heredity* (Alfred A. Knopf, New York, 1985); G. Searle, *Eugenics and politics in Britain 1900-14* (Noordhof, London, 1976). Greta Jones, *Social hygiene in twentieth century Britain* (Croom Helm, London, 1986); Kenneth Ludmerer, *Genetics and American society* (John Hopkins, Baltimore, 1972). On Germany, Sheila Weiss, 'Wilhelm Schallmayer and the logic of German eugenics, *Isis* vol. 77 (1986), pp. 33-46; Paul Weidling, 'Weimar eugenics: the Kaiser Wilhelm Institute for Anthropology. Human Heredity and Eugenics in Social Context'. *Annals of Science*, vol. 42 (1985), pp. 303-18. Bentley Glass, 'A hidden chapter of German eugenics between the two world wars', *Proceedings of the American Philosophical Society*, vol. 125, no. 5 (October 1981), pp. 357-67. On France, Linda Clark, *Social Darwinism in France* (University of Alabama Press, 1985). William Schneider, 'Towards the improvement of the human race: the history of eugenics in France', *Journal of Modern History*, vol. 54 (June 1982), pp. 268-91. On Russia, Loren R. Graham, 'Science and values: the eugenics movement in Germany and Russia in the 1920s', *American Historical Review*, vol. 82, no. 3 (1977).
4. Greta Jones, *Social hygiene in twentieth century Britain* (Croom Helm, London, 1986).
5. Elof Axel Carlson, *Genes, radiation and society: the life and work of H.J. Muller* (Cornell University Press, Ithaca NY, 1981).
6. E. Baur, F. Lenz and E. Fischer, *Human heredity* 3rd edition, 1927; translated and published in English (George Allen and Unwin, London, 1931).
7. Ibid., p. 41.
8. Ibid., p. 42.
9. Ibid., p. 182.
10. Review of *Erb Pathologie* (1937 Leipsig) by O. von Verschauer in *Journal of Heredity*, vol. 29 (1938), pp. 71-2.
11. Paul Popenoe, *Journal of Heredity*, vol. 25 (July 1934), p. 271.
12. Eugenics Society Papers, Eug. C58, Blacker to Carr Saunders, (8 February 1939).
13. *The Times* (17 January 1946).
14. B. Drzewieski, speech at Philadelphia, 25 March 1947, in *Reconstruction and Rehabilitation Newsletter*, vol. 1, no. 4 (April 1947), p. 1.
15. 'The race concept: results of an inquiry, UNESCO (1952), p. 6.

16. Ibid., Appendix 1. Statement on Race (issued 18 July 1950), p. 77.
17. Ibid., p. 78.
18. *Man* no. 220 (October 1950), p. 138.
19. 'The race concept' UNESCO, Fn. 15.
20. Ibid., p. 54.
21. Ibid., p. 27.
22. C.P. Blacker, *Eugenics Review*, vol. 42, No. 3 (October 1950), p. 22.
23. J.R. Baker, 'Freedom and authority in scientific publication', SFS Occasional Pamphlet, no. 15 (December 1953).
24. John H. Herz, *Political Science Quarterly*, vol. 63 (1948), p. 569: 'Detailed analysis of available facts and figures . . . substantiates what a second group of observers claims to be true, namely, that de-Nazification, which began with a bang, has since died with a whimper, that it opened the way toward renewed control of German public, social, economic and cultural life by forces which only partially and temporarily had been deprived of the influence they had exerted under the Nazi regime.'
25. 'The race concept', UNESCO, p. 30.
26. Ibid., p. 33.
27. Ibid., p. 32.
28. Hans Nachtsheim, 'Freedom and authority in scientific publication', Discussion at Congress for Cultural Freedom, Hamburg (1954), pp. 71–2.
29. 'The race concept' UNESCO, p. 35.
30. Nancy Stepan *The idea of race in science, Great Britain* (Macmillan, London 1982).
31. T.D. Stewart, *American Journal of Physical Anthropology*, NS, vol. 9, no. 2 (March 1951), p. 2.
32. 'Postgraduate training in physical anthropology', *American Journal of Physical Anthropology*, vol. 34 (1971, pp. 279–306.
33. See, for example, an article by Dr C.S. Stephenson, 'Blood Typing in American Samoa', *American Journal of Anthropology*, vol. 2, no. 2, July–September 1935, p. 233. Other articles on the genetics of blood groupings appeared regularly. According to Marcus Goldstein, 'In Recent trends in physical anthropology', the percentage of papers on genetics in the *Journal*, 1929–38, averaged 9.9 per cent, ibid., vol. 26 (1940), p. 193.
34. Ashley Montagu, 'A cursory examination of the relations between physical and social anthropology', *American Journal of Anthropology*, vol. 26 (1940).
35. Otto Klineberg, 'A study of psychological differences between racial and national groups in Europe', *Archives of Psychology*, no. 132, (ed. R.S. Woodworth), New York, 1931).
36. T.D. Stewart, *American Journal of Physical Anthropology*, vol. 34 (1974), Fn. 31, p. 2.
37. Burt Papers, D 191. Burt to his sister (3 August 1941).

5
The Arms Race and the Scientists

The Cold War brought about an increasing bitter political climate. In 1951, Eric Burhop wrote to Peierls about the libel case he was conducting against the *Daily Express*. The newspaper had suggested that the cancellation of his passport before a trip to East Germany was because he was planning to defect. The Atomic Scientists Association (ASA) was helping him in his legal action which was eventually successful. Burhop, however, felt the pressure acutely: 'Well, now that the Foreign Office has destroyed the privacy of our home life and done its best to tear my character to shreds, I trust I may be left in peace for a while'.[1]

Burhop was only one of many scientists in atomic physics who came under the scrutiny of the British and American authorities. In July 1947, the British government banned discussions of atomic science at the International Conference of Pure and Applied Chemistry. In the same year, it stopped a British Universities visit to Prague. British scientists found they were refused admission to the USA. In October 1952 the American *Bulletin of Atomic Scientists* listed the British scientists forbidden from entering the USA.[2] The Bulletin obviously felt that many of the decisions of the American government were taken on the flimsiest grounds. Polanyi recounted a particular bitter encounter with a US immigration official. Polanyi had listed his achievements in fighting Communism, explained his presence in the Society for the Defence of German Culture during the war (formed of mainly emigrés from the Nazis) and finally mentioned how Arthur Koestler's *Darkness at noon* had been dedicated to him. To his chagrin the official asked him who Koestler was.

However clumsy the American government's policy was in this respect, the British government felt increasingly that criticism of their lax security — in the context of the spy cases — was one of the major factors hindering exchange of atomic information between the two nations. It was therefore unsympathetic to scientists who were unorthodox on atomic policy. The aim was to convince the USA that Britain was a secure and dependable ally.

In October 1948, Sir Stafford Cripps had told James Forrestal, the American Secretary of Defence, that Britain 'must be regarded as the main base for the deployment of American air power and the chief offensive against Russia must be by air.' In a subsequent visit to Britain by Forrestal in November 1948 to secure American air bases on its soil, Prime Minister Attlee confirmed this. He also claimed, 'There was no division in the British public mind about the use of the atomic bomb — they were for its use. Even the Church in recent days had publicly taken this position.'[3] Attlee was referring to the pamphlet *The Church and the atom* published by the Church Commission in April 1948. In this pamphlet the manufacture, stock piling and use of atomic weapons, in the circumstances of 'defensive necessity' were justified. 'How easy it would have been,' commented the *Manchester Guardian*, 'to condemn atomic warfare outright and how irresponsible.'[4]

Judging by these reactions Attlee must have felt justified in his representation of public opinion to Forrestal. However not all scientists had given up the ghost. Several areas of Britain's nuclear policy came under criticism. One was the strategic value of a British atomic bomb.

What was the political background to this opposition? As many historians of this period have pointed out, in the period 1945–9 there were overwhelming political and strategic advantages for the USA in adopting a nuclear arms strategy.[5] First, atomic weapons permitted conventional forces to be reduced and the US government believed that the American people were unwilling to bear the burden of conventional forces in peace time. Secondly, the USSR was considered to have an invincible lead in manpower and other conventional forces but to be behind in the development of atomic weapons. Just how far behind was a matter for speculation. But overall the United States believed that a technological gap existed in nuclear weaponry. This gap and its maintenance were to assume enormous importance in the post-war period. This strategy depended less on any notion of a nuclear balance of terror than on the ability of nuclear weapons to act as a deterrent to a conventional attack by Soviet forces.

In these circumstances, what was the objective of the United States' sponsorship of the 'international control of atomic weaponry' and, in particular, of the Baruch Plan which the country put before the Atomic Energy Commission of the United Nations in June 1946? The Baruch Plan had a number of political advantages for the USA. It advocated a freeze on atomic weapons' development rather than the immediate abandonment and destruction of atomic weapons set out in the Soviet counter-proposals. This would have effectively maintained the position of the USA as the leading nuclear power in a world of non-nuclear countries. Attlee, for example, considered that American nuclear superiority meant the USA would never accept a plan for disarmament which involved the total abandonment of nuclear weapons. 'Renunciation of the weapon now puts any computation of strength of rival nations back into the pre-atomic bomb age. The US is unlikely to accept this in present conditions when power politics seem to be in full vigour.'[6] Moreover, the system of surveillance and inspection suggested by the Baruch Plan would have given the USA access to the military and civilian installations of the USSR. In an age of massive conventional air attacks, however, the policy of the USSR was precisely to disperse and conceal these installations. Further, the plan linked nuclear disarmament to a reduction in conventional forces — the one area of Soviet supremacy. Finally, it was to be enforced by the Atomic Energy Council, an organisation set up under the auspices of the United Nations. At that time, the United States had sufficient political control over the United Nations to ensure generally favourable attitudes to its policies. None the less, on these occasions the Soviet veto could be used. In the Baruch Plan the veto was to be abolished — again a major sticking point for the USSR. In these circumstances, it is unlikely that either power saw nuclear disarmament as likely. The USA would only countenance it on conditions which protected its political interests and especially its existing lead in nuclear weapons; the Soviet Union only on terms which admitted the country to equal status, or permitted it to advance towards it, and which did not 'freeze' the relationship in conditions of inequality.

The development of the British atomic bomb — exploded in October 1952 — was a different question. Many of the reasons the British government had for embarking on a policy of nuclear armament, are shrouded in secrecy. None the less, some factors are relatively well known. Britain saw the bomb as a guarantee of a continued place in the inner circle of international diplomacy. Between 1945 and 1949 this guarantee seemed necessary, not simply as a result of a Soviet 'threat', but also because of suspicion of

American indifference to British foreign policy objectives. Britain's atomic bomb was a means to assert British independence of the United States as well as to counter the USSR. In addition, the decision to build a British bomb was taken before Marshall Aid and the foundation of NATO. Before these events some doubt existed about American commitment to Europe. Moreover Britain, like the United States, believed a shift from conventional to atomic forces would, in the long run, both prove cheaper than maintaining a large conventional army and be more politically acceptable.

There is also evidence that, although the British government was gravely disappointed at the failure of the United States to continue co-operation and exchange of information in nuclear arms development in the post-war world, it hoped it could eventually change the American government's mind. It believed that it was possible for Britain to become a second rank nuclear power. Or at least that Britain would become nuclear at the same time or possibly before the USSR. According to Philip Oppenheimer in 1948, a year before the Soviet Union exploded an atomic bomb,

> One important factor may be the time necessary for the Soviet Union to carry out the programme of atomic energy to obtain a significant atomic armament. With all recognition of the need for caution in such predictions, I tend to believe that for a long time to come the Soviet Union will not have achieved this objective, not even the more minor but also dangerous possibility of conducting radiological warfare.[8]

A number of estimates made in the same year placed the explosion of an atomic bomb by the Soviet Union in 1952 or 1953. In fact, by then the Soviet Union had exploded a hydrogen bomb. Some scientists seemed to think that that fateful hour would not come for decades.[9] Because of this, British policy, though officially committed to Baruch, was not all that keen on it. In 1945, Attlee expressed doubts about the feasibility of international control. He was of the opinion that inspection of the sort envisaged by the Acheson-Lilienthal plan (a forerunner to Baruch) was impossible to carry out.[10]

There were of course a number of sources of instability in a strategy of this kind. The first came from the presumption that the technological gap could always be maintained at the same level; the second from the question of what conventional attacks could properly be regarded as justifying a nuclear response by the USA.

The third lay in the escalating political and military costs of nuclear strategy, especially one committed to a technological gap. These problems were particularly acute for Britain. However, they did not fully emerge until the 1950s.

Of all the scientists opposed to a British nuclear policy, Patrick Blackett devoted most thought to its strategic implications. He had a strong interest in military affairs. Blackett was born in 1897. During the First World War he was a naval officer. He became a Labour Party Fabian in the 1920s while at work in the Cavendish Laboratory. This occurred partly because of his friendship with Kingsley Martin, then editor of the *New Statesman*. Blackett's interest in strategic questions brought him into the Air Defence Committee in 1936 and from 1942 to 1945 he was director of operational research at the Admiralty. Blackett was involved with the development of the British A bomb from its earliest years. After the war he was appointed to the Scientific Advisory Committee whose function, before it became defunct in 1947, was to advise on the development of atomic power. He was therefore, in a good position to put his views on atomic armaments to the government.

However, by 1945 there were indications of the government's unwillingness to listen to what it considered were heterodox views on atomic armaments. In 1945, there were appeals for the internationalisation of atomic weapons to avoid the start of an arms race. Such views were heard even within the British Cabinet.[11] The government was also informed that there was an effort to mobilise scientific opinion behind internationalisation. In 1945, the government received a despatch from the British Embassy in Paris which described the efforts of Joliot-Curie and Paul Langevin along these lines.

> I have the honour to report that the internationalisation of atomic energy has become one of the main items of Communist propaganda in France. Monsieur Joliot-Curie the leading French scientist and Monsieur Paul Langevin, the doyen of the French scientific world, are both members of the Communist Party . . . Monsieur Joliot-Curie . . . said that the secret of the atomic bomb lay not in scientific but in technical formulae by which the Americans had worked out the production of the bomb. It had been acknowledged by a British scientist that any industrial nation could discover these formulae within six months. It was therefore not only criminal but pointless to try to preserve the secret.[12]

The reaction of those responsible in the British government for nuclear policy was to remove discussion of atomic development from the cabinet altogether. Henceforth an inner committee of the chosen and reliable decided upon atomic policy with little or no parliamentary or cabinet scrutiny.

Blackett's views on atomic weapons were, to a great extent, determined by his views on strategic bombing during the Second World War. He was then an ally of Sir Henry Tizard, who was one of the government's leading scientific adviser, and worked on the problems of air defence in the 1930s. Tizard and Blackett clashed with Lord Cherwell, then F.A. Lindemann, over the policy of saturation bombing of German cities, which they saw as futile. Tizard and Blackett lost the argument and the heavy bombing of German cities continued. Blackett, later, considered this policy had in fact prolonged the war. He believed that tactical air strikes against selected industrial and military targets would have been more successful in achieving a quick victory. Immediately after the war, Blackett saw the adoption of a nuclear arms strategy by Britain as a continuation of the Lindemann policy in that nuclear arms were weapons of massive destruction. He believed that this was an inhumane, inept, and ultimately futile military strategy.

Two memos which stated Blackett's objections were placed before the Scientific Advisory Committee in 1945 and 1947. They were received coldly by the government. They also led to Blackett's exclusion from future deliberation on nuclear arms policy, though he continued to act as an adviser to the government on other scientific matters. However, the aftermath of Blackett's exclusion was double-edged. Freed from government interference Blackett began to mobilise scientific support against a nuclear arms policy. In 1948 he published *Military and political consequences of atomic power*. This book was not an argument against having a defence policy for Britain: it was an argument against having a nuclear defence. First, Blackett believed that Britain, even in the atomic age, would have to continue to fight its wars by conventional means. The atomic bomb, like strategic bombing, would prove valueless as a means of achieving military victory. Moreover, Blackett foresaw a period of ten years elapsing before the intercontinental ballistic missile was developed. During this time there would have to be conventional back-up to nuclear attack. In addition, Blackett was less convinced than other nuclear strategists that the Russians would not catch up. He placed a Russian bomb five years away at the most. When this point was reached, the only function of nuclear weapons would be mutual annihilation. Ironically, in this state of affairs, conventional forces would once again come into their own. Consequently, Blackett proposed two things: the halt to a shift from a conventional to nuclear strategy, particularly in Britain's case, and greater efforts to bring about international nuclear disarmament. Therefore he pressed very hard for the Baruch Plan to be salvaged.

From the late 1940s until the late 1950s, Blackett continued to develop his theories but his audience contracted. By the late 1950s, however, various events revived interest in his ideas. The development of much more powerful hydrogen bombs and of intercontinental ballistic missiles projected nations into the era of massive nuclear retaliation. Out of this grew the deterrent theory of nuclear weapons. However, nuclear deterrance implied nuclear weapons were a 'last resort' weapon. What then would happen if the Soviets launched a massive conventional attack? How would wars and disturbances be effectively handled short of massive mutual destruction? What did the development of a range of medium-size (by the standard of atomic and hydrogen bombs) weapons imply for battlefield tactics?

Blackett was one of the few thinkers on military strategy who had continued to assume that conventional forces would still be deployed in a nuclear age. He had also fully grasped the idea that the existence of nuclear weapons led to a whole series of new problems in military strategy. Blackett's views were increasingly sought, particularly after 1957 when Britain's vulnerability in nuclear policy was revealed. He was invited to give lectures at the Imperial Defence College in 1959, and at the NATO Defence College in Paris in the same year. He was also on the executive committee for the Institute of Strategic Studies and his views were widely canvassed by, for example, Henry Kissinger then making a name for himself as an expert on nuclear strategy.

In 1954, Blackett believed that the USSR had demonstrated the ability to catch up with the USA and that, given this nuclear counterbalance, conventional warfare would once again become important:

> Thus since 1949 we have been watching the value of the main ingredient in our national defence arsenal gradually diminish as the Russians build towards a stockpile of atomic bombs which they will feel . . . will some day reach sufficient proportion to cancel out the atom as an instrument of warfare. If such an impasse occurs, the United States would appear

to be left in a rather unenviable position. The most useful product of our technological competence would appear to be lost to us except as a deterrent to the use of A bombs by the enemy and the Russians would appear to be free to take full advantage, in world military and diplomatic affairs, of their vast superiority in manpower and their highly strategic position dominating the Eurasian land masses.[13]

For a while Blackett became an advocate of limited nuclear war. This would rely on a combination of conventional forces and tactical nuclear weapons. He pushed this against the theory, current in some circles in this period, of massive nuclear retaliation in the event of any Soviet incursion, conventional or nuclear.[14] Much of the interest in Blackett's ideas was provoked by this concept. But Blackett's study of atomic strategy, begun in the 1940s, had been subordinate to his perception for the need for nuclear disarmament. By 1958, nuclear disarmament was gradually reasserting itself in his work. Thus in the *New Statesman* of 17th May 1958 he turned against the idea of a limited war fought with tactical nuclear weapons.[15] This could not, he considered, be limited in any real sense of the word. Nor was it possible to keep any nuclear war limited. Further, Blackett replied in this vein to a query from Mountbatten about whether a recent lecture of his had advocated renunciation of tactical nuclear weapons in any future conflict and a shift to conventional forces in NATO:

> The key point seems to me more and more to be whether the West would initiate the tactical use of atomic weapons if the Russians did not. I believe that to do so would hasten defeat if the West had numerical inferiority of ground troops. What is the way of this dilemma I am not sure.[16]

From 1945, Blackett shared with a minority of other scientists a belief in the importance of international control. By 1948, these scientists had virtually written off the Atomic Scientists Association as a vehicle for applying political pressure for this aim. After the disastrous meeting of 1948, in which the issue of Lysenko had taken over from that of atomic disarmament, Blackett wrote to Eric Burhop,

> I am sorry the ASA meeting went so badly in the afternoon — I left after Peierl's speech as I was dog tired. I must say I

feared the ASA would go the way it did. It is asking for trouble in my view to put up general speakers on anything connected with Russia. I have not any very firm views about the future of the ASA. Probably the best thing is to keep it in being and prevent it being too flagrantly misused.[17]

However, the Association of Scientific Workers was a different proposition. In the late 40s, the AScW had around 15,000 members mostly in industrial or government scientific occupations. Unlike the ASA which was largely a body of elite scientists, many with close government connections, the AScW was much more representative of the scientific rank and file. It tended, in addition, by its very nature to represent the trades-union-minded among scientists. Also, as a trades union, it had connections with other organisations in the Labour movement.[18]

In 1947, the AScW issued a political statement on the future of the Baruch Plan. It conceded that the real problem of getting international control of nuclear arms was the political tension between the USSR and the West and that a water-tight solution to control could never be found. However, it emphasised the wide area in which agreement existed and suggested a series of compromises on contentious issues. The AScW suggested that the Soviet Union's demand for a ban on the use of atomic weapons be met, while reserving the right to their use by the United Nations in any general disarmament proposal. The Association stated that,

> The atomic bomb is a weapon of indiscriminate mass destruction and so cannot be considered as a policing weapon by an international security force. The only possible moral justification for its use in the future would be as a deterrent, so that any nation that itself first used atomic weapons would know that swift and sure reprisal in the same form would be invoked against it.[19]

The AScW also suggested a compromise on the question of dismantling existing stocks of nuclear weapons. The United Nations was to take over the US stocks provisionally, pending the time when the USSR would fall in with the provisions of a general disarmament agreement. The AScW strongly concurred in the Soviet opinion that an international agency should largely be concerned with weaponry and not with nuclear energy as such, whose development they welcomed:

> In the future, atomic energy may become a key factor in the economic life of nations. The economic power that ownership or control of atomic energy plants will confer could enable economic pressure to be exerted for the purpose of forcing a change in the economic structure of whole nations. Spokesmen of the USA have never hidden their desire to maintain or restore a system of 'free private enterprise' over as large a portion of the earth's surface as possible. Socialist Russia is naturally anxious that any international system of the control of atomic energy should not make possible the use of the economic power of atomic energy for this purpose. We are of the opinion that a Britain committed to a planned Socialist economy could be quite entitled to similar misgivings.[20]

The AScW suggested further compromises on the question of inspection to meet the objections of both the USSR and USA. It came out in full support of the USSR on the question of the veto.

In *The atom and the charter* (*Science and Social Affairs* no. 1, published by the AScW and the Fabian Society, London, 1946), Blackett discussed the veto. He argued that the USSR could not be expected to put the question of atomic weapons to the United Nations Security Council since the country had been, and was likely to remain, in a minority position on that body. Also, the enforcement of 'sanctions' decided upon by a Security Council majority against either USSR or USA would involve, in either case, a major global war — precisely the situation which disarmament was intended to avoid. By implication Blackett said, the only sanction against transgressors of agreement was the breakdown of that agreement and the return to an arms race and international tension — a situation already in existence. Thus, Blackett seemed to be saying, agreement was ultimately dependent on a real desire for it on behalf of both parties and perception of the disasters consequent on its absence. Concord had to be based on the highest level of mutual agreement.

The AScW saw the British government as being in a position to use its influence to bring about agreement, but it felt:

> Very disappointed that the role taken by the United Kingdom in the discussions on the control of atomic energy had been almost entirely negative. The attitude of our delegate to the Security Council has consisted far too often in merely saying 'me too' to any proposals put forward by the United States delegate.[21]

The chances of influencing government were slim. However, there were other channels through which the AScW could direct its proposals. Since 1942, the AScW had been affiliated to the Trades Union Congress (TUC) and in 1945 it had created, jointly with the TUC, a Scientific Advisory Committee. During the years of his presidency of the AScW, Blackett spoke at the annual conferences of the TUC on several matters concerning science, in particular (along with the Chemical Workers' Union) on radiological protection for workers in the nuclear industry and on the need for freer access to scientific information. He also moved a resolution in 1946, on behalf of the AScW, asking the government to renounce the use of the atomic bomb and stop its manufacture. From 1946 until the mid-50s, when C.F. Powell also moved a similar resolution as president of the AScW, the AScW was behind or associated with most of the resolutions on the banning of atomic weapons before the TUC. But whereas the TUC was sympathetic to the more exclusively industrial problems brought before them by the AScW, the resolutions on the bomb fell on stony ground. Most resolutions were remitted to the General Council of the TUC for study — that is shelved.

In fact, between 1945 and 1949, the growth of cold-war sentiment blocked off avenues of discussion on these topics. In the case of the TUC, the signal for this came with its departure from the World Federation of Trades Unions (WFTU) in 1948 and affiliation to the International Confederation of Free Trades Unions (ICFTU) in 1949. The break with the WFTU had been delayed on the advice of the British government, largely because it was hoped that the British delegation might exercise sufficient influence to swing the WFTU behind the Marshall Plan. When this failed, the way was open for the separation of the international trades union movement into two wings: the one organised by the WFTU and dominated by the trades union movements in Communist countries and by countries in which large Communist parties existed; and the other the ICFTU, largely dominated by the American trades union movement together with those European countries and their colonies which were firmly in the American alliance.

The separation had numerous political consequences. On the question of the atomic bomb, it meant the end to the 'truce' and the practice of remitting resolutions on disarmament to the TUC General Council. Instead the General Council followed a persistent policy of swinging disarmament proposals behind the Baruch Plan and, even more revealingly, keeping all motions close to current disarmament policy of the United States. The Trades Union Congress of 1950 also warned against the 'Peace Campaign':

> The 'Peace Campaign', in demanding the banning of atomic bombs, deliberately ignores the action of the Soviet Union in blocking proposals formulated on behalf of the United Nations to place under international control the raw materials for research and productive equipment required for the peacetime use of atomic energy, and for adequate inspection to prevent secret production of atomic weapons along with the assurance of America's readiness to surrender the atomic bomb to a world authority.[22]

Even an anodyne motion on atomic weapons in 1950 from W. Padley of the Union of Shop, Distributive and Allied Workers (USDAW), regarded by that union as supplementing rather than contradicting the General Council's resolution on international relations, was treated with suspicion. Much to his surprise, Padley found himself the subject of pressure to make his resolution into an amendment and hence not only to curtail the time available for speeches for and against it but also to have it treated as a direct opposition. Sir Vincent Tewson, the TUC president, insisted that, 'The General Council are particularly anxious that there should not be two voices and two decisions on any question so vital as this.'[23]

In addition, the AScW found 'middle opinion' notably unsympathetic. This unsympathetic response came from some who were later prominent in the Campaign for Nuclear Disarmament (CND) but who, in the 1940s, were gripped by cold-war fever. On 22nd December 1947, the *Manchester Guardian* reported that a group of prominent opinion formers, including Earl Russell, T.S. Eliot, Lord Brabazon, Lord Quibell, Lord Vansittart, the Reverend Gordon Lang, Tom O'Brien MP, Raymond Blackburn MP, and Clement Davies MP, had issued a statement in which they pointed out:

> The respects in which Communism, Fascism and Nazism resemble each other, and the concern felt at the growing strength of Communism (so that) all attempts to control atomic energy have been frustrated by the intransigence of the Soviet Union . . . If this last and most powerful intervention were to fail we suggest that the freedom-loving powers . . . should act in concert. They should henceforth develop

such a predominance of defensive strength including atomic strength that no power would be able to challenge them.

The statement went on to call for religious, moral and spiritual consciousness.[24] As far as 'middle class' radicalism was concerned its adherents were scattered and ineffective until the upsurge of the middle 50s. Its representatives were on the whole either ignorant of the significance of atomic weapons or won over to a warlike posture towards the USSR. Only a few groups, either pacifists, Quakers or representatives of the pre-war peace organisations were prepared to challenge the British government's nuclear arms policy. These peace organisations included the Peace Pledge Union and the National Peace Council. Even among these groups there were representatives of science. One group which combined the pacifism of the Quakers and the internationalism of the Peace Pledge Union was the Medical Association for the Prevention of War (MAPW) founded in 1951, in which Dr Richard Doll played an active part. Its vice-presidents included John Boyd Orr, Professor J.M. Mackintosh, Dr H. Joule, Dr B. Stross and W.C.W. Nixon. Kathleen Lonsdale and Professor L.S. Penrose, both Quakers, were also members.[25]

The objects of the MAPW were to consider the ethical responsibility of doctors in the event of war, to oppose the use of medicine for the prosecution of war, and to divert defence expenditure to the fight against world hunger. It also promoted international co-operation between doctors, sought to make contact with Soviet counterparts, and was interested in studying the causes of war, particularly its psychological roots.

This group was born from a combination of the pressing problems arising from the post-war increase in international tension with the older pacifist and internationalist tradition in British politics. The MAPW contained a high proportion of Quaker and 'progressive' opinion. For example, it was affiliated to the National Peace Council founded in 1908, which contained such representatives of centre-left and internationalist opinion as Richard Acland (formerly of the Commonwealth party), Vera Brittain and Rita Hinden of the Fabian Research Bureau.

The liberalism of the MAPW was both a strength and a weakness. It was a strength because it signalled the deep roots of the MAPW in the British liberal political tradition. These were strong roots indeed, for in 1951 the political centre was fast disappearing. But it limited the response of the MAPW to post-war problems. Whereas the AScW offered political and social analysis of the crisis in international relations, the MAPW looked at human nature. The theories of the psychologist William Trotter proved particularly alluring. Trotter, a liberal, had explained the outbreak of war in 1914 as due to the re-emergence of the agressive instincts implanted in the human race in early evolution. In particular he emphasised how war-fever led to the enforcement of conformity, the persecution of dissent and the prevalence of panic. This analysis might have seemed particularly appropriate to the liberals in the MAPW in view of the political atmosphere in 1951.[26] But the MAPW, like the other small middle-class internationalist groups, had little which was significant to say about the specific problems of atomic war. Good as its intentions were, it showed the limitations of these groups.

In contrast, the AScW, was in a much better position to produce telling criticisms of government nuclear policy. In 1950, as a riposte to the government's *Manual of basic training on civil defence*,[27] it produced a pamphlet called *Atomic attack, can Britain be defended?* In the foreword, Blackett set out his beliefs on the civil defence question:

> I think the arguments presented here make inescapable the conclusion that adequate defence of the United Kingdom against atomic attack launched from nearby bases on the Continent is quite impracticable, if only on economic grounds. The cost of the necessary passive defence measures, dispersal, underground factories, shelters, etc is wildly outside the economic possibilities of this economically hard-pressed country. It is entirely safe to conclude that these steps will not in fact be taken. This conclusion rests on the peculiar geographical position of the United Kingdom and on the high congestion of its population and in no way conflicts with the conclusion that atomic bombs alone are not likely to be quickly decisive in a major war between Continental powers such as America and Russia.[28]

This brought down on Blackett the wrath of Sir John Anderson, former chairman of the government's advisory committee on atomic energy, who criticised the pamphlet in the *Evening Standard* of 19th June 1950.[29] Anderson was a scientist who had had a long and distinguished career in public service. In the late 1930s he was in charge of Britain's civil defence programme and, in the Second

World War, one of his responsibilities was Britain's secret defence work including the development of the atomic bomb.

Anderson's target was Blackett's view of the indefensibility of Britain in the event of atomic attack. Anderson considered the cost of atomic weapons would limit their supply and that the dangers of atomic radiation were exaggerated. Above all he believed that a policy of deep shelters could be adapted relatively easily to protect the civilian population against atomic attack. But Blackett stuck to his views and, in the end, these were not very different from the government's own conclusions in the White Paper of 1957, that Britain was indefensible in the event of an all out nuclear war. But in the early 1950s the government found this an embarrassing admission and it publicly resisted making it. This was especially so since the AScW had some success with its attack on civil defence. There was considerable demand for its pamphlet, which went into two editions, and in March 1950 the Sheffield Branch of the AScW organised a conference at which these questions were discussed by 300 delegates from various parts of the labour movement and from professional organisations.[30]

The efforts of AScW were augmented by those of the World Federation of Scientific Workers (WFSW), founded in 1946. However, this grouping had more difficult obstacles to overcome. For example, in January 1951 the General Council of the TUC warned its members against the efforts of the WFSW to organise the Science for Peace Conference. A distinctive characteristic of the WFSW was that it was not split along East-West lines. An organisation of scientists of this sort immediately roused suspicion and this was compounded by the fact that its leadership, like that of the AScW, tended to the left and, in some cases (Bernal for example) was overtly pro-Soviet. In addition, the WFTU, to which the AScW was affiliated, was on the government blacklist as a 'suspect' organisation. This severely limited its influence. Those who were not for the West in 1948 were considered to be against it. Nonetheless, in the succeeding years the WFSW was able to exercise a good deal of influence, in a backstairs way.

The foundation of the WFSW led, under the aegis of Bernal, to the development in Britain of a National Committee of Science for Peace whose membership also included J.B.S. Haldane and Burhop. The Committee held a series of conferences on the need for atomic disarmament. The participants included Max Born, the German physicist, John Boyd Orr, the nutritionist, and Kathleen Lonsdale. The presence of these three showed the potential of

Science for Peace for attracting the non-Communist left. But the potential was never realised. Although they subsequently held a conference in January 1952 attended by 180, and kept plugging away at the public and scientific conscience, Science for Peace never achieved its aim of mobilising a broad section of liberal and left opinion against nuclear weapons.[31] Its leading members found difficulty in travelling; many of the participants to its conference were refused entry into Britain.

In the early 50s a rigorous conformity, often self-imposed, had settled over the scientific community. Nor were nuclear weapons a major political issue until March 1954, when there was a dramatic change in the climate in which nuclear weapons were discussed.

Notes

1. Peierls Papers, F1-11, Burhop to Peierls (27 August 1951).
2. *Bulletin of Atomic Scientists*, vol. 8, no. 7 (October 1952).
3. *The Forrestal diaries*, Walter Millis (ed.) (Cassel and Co., London, 1952), p. 460.
4. *Manchester Guardian* (16 April 1948).
5. C.J. Bartlett, *The long retreat: a short history of British defence policy, 1945-70* (Macmillan, London, 1972); C. Driver, *The disarmers: a study in protest* (Hodder and Stoughton, London, 1964). E. Meehan, *The British left wing and foreign policy* (Rutgers University Press, New Brunswick, 1960). A.J.R. Groom, *British thinking about nuclear weapons* (F. Pinter, London, 1974); and Robert Gilpin, *American scientists and nuclear weapons policy* (Princeton University Press, New Jersey, 1962).
6. Attlee Cabinet, CAB 129/4 Memo CP (45) 272 (5 November 1945) (Public Record Office, Kew).
7. The planning of the British atomic programme was entrusted to a small committee of a few chosen Cabinet members and the Prime Minister. Many of the papers of this committee are still unavailable at the Public Record Office, Kew. Discussions of atomic energy and arms in the full cabinet are very perfunctory until 1954.
8. Gilpin, *American Scientists*, p. 75. See also A. Cole of the Atomic Energy Commission. 'The Soviet Hydrogen Test occurred sooner, than most officials in Washington had expected . . .', *Atomic Scientists Journal*, vol. 3, no. 3 (January 1954), p. 160.
9. Bartlett, *Long Retreat*, Fn. 5, p. 33.
10. Attlee CAB 129/4, Fn. 6 (5 November 1945).
11. CAB 128/4, CM (45) 51st Conclusions, Cabinet Office, (8 November 1945) (Public Record Office, Kew).
12. CAB 130/8 Gen. 106-114, No. 107 (13 December 1945). Atomic Energy in France, Despatch from the British Embassy in Paris (Public

Record Office, Kew), pp. 1-2.
13. 'America's atomic dilemma', *New Statesman* (13 February 1954), pp. 180-2.
14. See the pamphlet written for the Royal Institute of Strategic Affairs with Sir Anthony Buzzard, Denis Healey MP, and Richard Goold-Adams MP, 'On limiting atomic war', *Bulletin of Atomic Scientists*, vol. 13, No. 6 (6 June 1957), p. 216.
15. 'Limited nuclear war', *New Statesman* (17 May 1958), pp. 625-7.
16. Blackett Papers, F81, Blackett to Mountbatten (18 August 1959).
17. Ibid., J17, Blackett to Burhop (12 November 1948).
18. *The Scientific Worker*, vol. 5, no. 4 (July 1950), pp. 23 -7; P.M.S. Blackett, 'The development of the Association of Scientific Workers' (1947), and 'Summary of the presidential address to the AScW', (25 May 1947), unpublished manuscripts in the Blackett Papers, E23 and E22.
19. 'The international control of atomic energy', Statement of the executive committee of the AScW (August 1947), p. 2.
20. Ibid., p. 3.
21. Ibid., p. 4.
22. Reports of the Proceedings of the 82nd Trades Union Congress, Brighton (1950), p. 197.
23. Ibid., p. 409.
24. *Manchester Guardian* (22 December 1947). For Russell's equivocation on the bomb in the 1940s — at one stage he seemed to be advocating a preventive nuclear strike against the USSR as a means of securing her agreement to the international control of atomic energy — see Ronald W. Clark, *Life of Bertrand Russell* (Cape, London, 1975), pp. 525-69. Russell was certainly never pro-nuclear, but in the 1940s he accepted the British government's line on the international control of atomic weapons. See also House of Lords Debates (30 April 1947), vol. 147, pp. 272-3.
25. Penrose Papers, 41/3, Memoranda and Leaflets; and 41/4, correspondence of MAPW.
26. The MAPW were particularly interested in the psychological theories of William Trotter. For a discussion of Trotter, see Greta Jones, *Social Darwinism and English thought* (Harvester, Brighton, 1980).
27. *Manual of basic training*, Civil Defence II (HMSO, London, 1950).
28. *Atomic attack: can Britain be defended?* (Association of Scientific Workers, London, 1950), p. 1.
29. *Evening Standard* (19 June 1950); see also Blackett Papers, F54 (1950); for a fuller account of Anderson's career, see the *Dictionary of national biography* (Oxford University Press, 1971), pp. 21-4.
30. *The Scientific Worker*, vol. 5, no. 4 (July 1950), 32. The Churches sent 21 people. The Amalgamated Union of Engineering Workers and National Council of Labour Colleges also participated. The AScW claimed that on the publication of their 1947 statement their North-Eastern Area wrote to all trades unions, co-operative guilds and churches in the South Yorkshire area offering to provide speakers and enclosing copies of the report. The Sheffield branch of the AScW was particularly active. The dissemination of AScW material to trade unions and cooperative branches went on throughout the 1950s.
31. See 'Science for Peace', *Nature*, vol. 169 (15 March 1952), pp. 499-50.

6
Towards a Nuclear-free World

In 1952 the United States conducted a series of tests of thermonuclear devices, which were steps in the development of a hydrogen bomb. In August 1953, in spite of some initial scepticism, the Soviet Union was conceded also to have exploded a hydrogen bomb. In early 1954 the United States embarked upon a further series of tests on Bikini Atoll in the Pacific and on 1st March 1954 conducted the Bravo test shot which it routinely announced to the world's press. However, on 14th March 1954, a Japanese fishing boat, the *Lucky Dragon* (*Fukuryu Maru*), docked at its home port of Yaizu. It had been fishing in the vicinity of the Bravo test shot although outside the area designated as unsafe by the United States government. On board the *Lucky Dragon* were a number of sick men of whom one eventually died. These men told the story of a fine white ash falling on the boat minutes after the explosion some 70-90 miles away. A few days later, symptoms of sickness appeared in members of the crew and the boat ran for port. On docking, the cargo of fish was sold and dispersed throughout Japan. Nine other fishing boats in the area were affected by radioactive fall-out from the Bravo test shot, but to a lesser extent, as was a US naval vessel, the *Patapsco*. In addition, several islands under United States jurisdiction 330 miles away were affected and subsequently evacuated.[1] The chairman of the Atomic Energy Commisson Lewis Strauss, admitted on 31st March 1954,

> The radioactive yield was about double that of the calculated estimate — a margin of error not incompatible with a totally new weapon . . . The shot was fired. The wind failed to follow the predictions but shifted south of that line and the little islands of Rongelap, Rongerik and Uterik were in the path

of the fallout . . . The 23 members of the *Fukuryu Maru*, 26 American personnel manning weather stations on the little islands, and 236 natives of these islands were therefore within the area of the fallout.[2]

The test shot programmes by both Russia and America — to say nothing of the United Kingdom — were surrounded by intense security, and it is unlikely that the world would have heard of this miscalculation if it had not affected the hapless Japanese boat and then been broken to the world through the Japanese press. It was given particularly prominent coverage in the non-aligned press. The Indian government protested publicly at the incident. But the effect in Britain and America was also startling. Gone was the easy-going tone with which British newspapers greeted Britain's atomic test programme in 1953. At that time the *Manchester Guardian* correspondent was dismissive on the question of fall-out,

> An atomic explosion in the heart of Australia of the type now projected could hardly present the least danger if the sky was clear, regardless of the direction of the wind. But if radioactive particles wafted southwards or eastwards they would doubtless affect camera plates in big Australian capitals and provide harmless but sensational evidence of a nuclear discharge capable of dramatisation. The sequel would be protests and alarmist speculation by irresponsible scaremongers such as the authorities most dread for psychological reasons.[3]

In 1945, in the wake of the atomic explosions at Hiroshima and Nagasaki, a strange phenomenon was noted by photographers worldwide: a fine film of radioactive particles was deposited upon their photographic plates. This was the first public demonstration of 'fall-out' from nuclear explosions and, initially at least, it was seen in the light of a curious though harmless incident.[4]

Radioactivity as such had been known about since the turn of the century. It was produced by the emission of particles (or 'decay') of substances such as radium. Each substance had a 'decay' rate or rate of emission. In some cases the decay rate was a short span of time. In others, such as plutonium, it extended over thousands of years. The ability of the emissions and of X-rays to cause chemical changes in photographic emulsions gave rise to radiography and, in the twentieth century, this became an important adjunct to medical science. But it was also observed that radiographers could be subject to injury or sickness. In 1927, H.J. Muller demonstrated an invisible hazard. He showed that genetic damage could be produced in mice by exposing them to X-rays. He concluded that similar damage was possible in humans.[5]

However, the main immediate concern was the health of radiological workers and to deal with this an International Radiological Protection Board was set up. Similar boards were also established in various countries whose industries and health institutions made use of radioactive substances — such as luminous paint, for example. These devised a set of standards for the protection of radiological workers.[6]

When the atomic energy industry was created in Britain, the health of workers at Harwell became a concern. The Chemical Workers Union, which covered workers in industries using radioactive materials (including those in the atomic industry) and the AScW (whose members also worked in atomic energy), joined together as a pressure group for better radiological protection and the MRC set up a Tolerance Doses Panel to deal with these questions. The focus was almost entirely on possible industrial hazards. The long term genetic effects were not a prime consideration at this stage.

The military use of atomic energy caused a major restructuring of thought on this question. A major source of information about radioactivity arose from the atomic weapons test programme of the USA but since this was conducted under military (in fact Navy) direction, information was severely restricted. For example an article published in *Science* in 1953 by scientists working under military authority claimed that whilst, 'during the past 20 months there has been an increase in the tempo of atomic weapons tests leading to an increase in atmospheric radiation',[7] this was only of interest, according to this article, to certain restricted groups: scientists engaged in low level radiation measurement (who could, using simple apparatus, measure the increased radiation caused by atomic explosions); uranium prospectors, whose geiger counters were presumably causing waves of false optimism during a test explosion; and the photographic industry, whose photographic plates were affected.

Two sorts of bombs were created by the atomic or nuclear arms programme: from 1945 to 1952, atomic or fission bombs, whose explosive power was generated by splitting atoms of heavy elements like uranium and plutonium creating a chain reaction. From 1952 work was underway on thermonuclear or fusion bombs, which were made by the joining of isotopes of hydrogen ignited by means of an atomic explosion. Most early experiments in this produced devices rather than deliverable bombs. Both processes produced radioactive fall-out but the difference was that hydrogen bombs were more powerful; radioactivity from hydrogen bombs penetrated to the stratosphere leading to global distribution of fallout. The possibility of creating hydrogen bombs which were or could be 50 times more powerful than atomic or fission bombs worried the American scientist Robert J. Oppenheimer, whose career in the American nuclear arms programme subsequently floundered on his lack of enthusiasm for the hydrogen bomb programme.

Evidence about radiactive fall-out from explosions was, however, kept from the public. Writing in May 1952, Dr W.G. Marley, an official in charge of civil defence in Britain, produced a scathing conclusion about public concern. He believed radioactivity, 'Loomed rather larger in the public eye than was really justified when the causes of the casualties are examined . . . This was perhaps due to the mystery attached to the subject of radioactivity and also possibly to the memory in people's minds of the injury sustained by early workers, 40 years ago, in the field of X-rays.'[8] According to Marley, the main civil defence problem was that of civilian 'morale'.

A secret and less reassuring history was going on behind the scenes. According to the Australian Royal Commission on the atomic explosions conducted by Britain in Australia in the 1950s, even before the Bravo test shot there was miscalculation and ignorance about the effects of fall-out. The Commission estimated that Token One, the British explosion of October 1952, was subject to an under-estimation of explosive yield. The British authorities estimated it as 5 kilotons, whereas it was 10. Token One was conducted in the wrong weather conditions leading to the failure to disperse of the radioactive cloud it produced. It subsequently moved intact across country. In addition, the fall-out was underestimated by three times, which meant that the zone subject to risk was seriously underestimated.[9]

To be fair, it took time for governments to collect and evaluate the consequences of nuclear weapons. For example, an Atomic Bomb Casualty Commission set up to examine the effect of the Nagasaki and Hiroshima explosions concentrated primarily on immediate effects. The long-term consequences were much slower to emerge and Dr Marley could state fairly in 1952 that only 5.15 per cent of Japanese casualties at these explosions were caused by radiation. In British cities, whose buildings had brick walls, he estimated this rate would be reduced to four per cent.[10]

What then did scientists and especially geneticists think? The MRC Tolerance Doses Committee co-opted geneticists onto it. As early as the 1940s it was concerned with the question of the effect of increased genetic mutation rates on the population. L.S. Penrose had attempted some estimate of this by using records of radiological workers but they were in such chaos it proved impossible. J.F. Loutit, who was responsible from 1948 for conducting these enquiries (by 1957 he was in charge of the MRC's Radiological Research Unit at Harwell) wrote to Penrose on this issue. The reply he received was reassuring: 'Even if some of the mutation rates are raised we have no reason to suppose this is necessarily a bad thing in a wild population like the human species.'[11] By this Penrose meant that a small increase in mutation would be unimportant if not selectively bred for in a population. However the MRC admitted its ignorance

> The whole question of the genetic aspects of tolerance to radioactivity is still very much in the melting pot. The protection sub-committee of the MRC could not get any general agreement among itself. Haldane, for one, held very strong views and the matter now reads that the protection sub-committee has recommended that the MRC set up a special panel of geneticists to consider this problem. In fact the whole structure of the MRC committees on atomic energy is undergoing revision. Some day therefore, the pot will be put on to boil.[12]

The scientists working for government were aware by 1950 that, in addition to workers in nuclear industry, the general public could expect a general rise in background radiation. They were aware of the effect of this on mutation rates. Loutit wrote to Penrose that 'Some members of the Panel were particularly anxious to know how far the spontaneous mutation rate accounted for miscarriages, still births and neo-natal deaths, or putting the question another way: given the stillbirth factors and neo-natal rate etc. as it is at present, if one doubled the natural background how would these vital statistics be altered?'[13] This was of course important for it represented a concrete statistical demonstration of the deleterious effects of radiation, and would be of concern to the general public.

Between 1945 and 1963 there were at least a total of 362 atmospheric tests[14] but the crucial year was 1954. The Bravo test shot incident, more than any other of the 1950s, revealed the problems of fission and particularly hydrogen bomb explosions. The advent of thermonuclear (H bomb) explosions led to stratospheric pollution and, consequently, the distribution of radioactivity over the whole earth. The Bravo test shot demonstrated the physical effects of radiation and how it was dangerous to persons not within the actual explosion area — a concept which had been hazy in the minds of the public acquainted chiefly with the effects of Second World War bombing. Further, it showed the extent to which nuclear war was, by its nature, a global phenomenon; a fact underlined by the reports which subsequently flowed in recording higher levels of radioactivity, as a result of the Bravo test, in widely dispersed areas throughout the world.

The substantial disquiet felt at this turn of events, which made headline news in the press, had a number of long-term effects. The concern was felt at the very highest level of government. Winston Churchill, for example asked Lord Cherwell to comment on these events. What should the public be told? Cherwell concluded in a memo for Churchill on fall-out in November 1954,

> It seems to me very important to concert with the US whether the public should be told anything about the probable results of H bombing and if so what. If our ideas on the subject are right the effects might be so catastrophic as to deter even the Russians from risking a war. On the other hand publicity could cause great anxiety here and possibly panic in America.
>
> A certain amount is of course already known or can be guessed from all the talk about the Japanese fishermen, so that it is doubtful whether a policy of permanent silence is feasible even if it were desirable . . . The question is of course closely linked up with Civil Defence. If the 'fall-out' is so important as is now believed all our plans for this will have to be reconsidered. . .
>
> . . . On the other hand it would be unfortunate if the public formed the idea that the authorities did not know what they were about — which might happen if the question were not carefully handled.
>
> Everything naturally depends on whether and when we think the Russians will be in a position to use H bombs. All these years explosions have been of the very small A bomb

nature, and I am inclined to think they have not yet got A bombs which can be carried in an airplane, though they will no doubt develop this in the course of two or three years.[15]

Cherwell suggested a senior minister to be sent to Washington to discuss these questions.

In fact the government's policies were in considerable disarray. In a series of Cabinet discussions which took place in 1954, it was admitted that the government shared the widespread public surprise at the extent of radioactive contamination following the explosion of 1st March. One minister described the situation as 'sombre and awe-inspiring'. The government's civil defence plans were rapidly revised although the government was worried about admitting this publicly. A memorandum by the Minister of Defence in December 1954 described their fears:

> It is, I think, evident that this new information must have a revolutionary effect over a wide range of our war plans, both military and civil. Thought is already being given to its implications by the limited circle of Ministers and officials to whom the scientific appreciation is known. But we cannot ensure that all our preparedness will be properly adjusted to allow for this new factor without widening the limit within which knowledge of the new implications has so far been confined. Unless this is arranged much of our planning is bound to get out of gear.
>
> If this is done, however, we must accept some risk that people may come to know quite soon that the government are planning on this new hypothesis . . . Much of the present indifference of the public would vanish if they found out that the government had adopted this basis for their defence plans.[16]

The question of public reaction was raised again at a subsequent meeting. Informed of the BBC's plans to broadcast a programme on the hydrogen bomb, the Cabinet felt that some control over its content was necessary.

> The chairman of the Atomic Energy Authority had spoken about this to the director-general of the BBC, who had undertaken to make himself personally responsible for ensuring that those planning the programme consulted reputable scientists

(who would be in touch with scientists employed by the Authority) and that the programme would be free of political bias. Was it enough to leave this to the discretion of the director-general of the BBC?

It was the general view of the Cabinet that further guidance should be given to the BBC by a responsible Minister. It was important . . . that the government should themselves retain control over the form and timing of publicity on the effects of thermonuclear weapons.[17]

But public opinion was already becomng alerted to the nuclear fall-out issue, thanks to the *Lucky Dragon*. The best known incident was the refusal on 6th April 1954 of the Labour-held Coventry City Council to vote funds for civil defence (32 votes to 13). The leader of the Council explained this as a means 'to strengthen the hands of those who want to outlaw the bomb.'[18] In other areas of Labour strength, including Rotherham, Salford, Derby, Hertfordshire, Birmingham, Bethnal Green, Ipswich, Newcastle, Dagenham, Glamorgan, Birkenhead, Fife and South Shields, there were similar, though unsuccessful, moves to block civil defence expenditure.[19]

The character of the protest at local level can be illustrated by events in one Yorkshire district. On 24th April 1954, the Rotherham Trades Council called for a ban on the H-bomb at its monthly meeting. The resolution in support of this (opposed by two delegates) was passed to the TUC. Three days later, on 27th April, the Rotherham Labour Party put forward a resolution sponsored by the Amalgamated Engineering Union (AEU) No4 Branch calling for a British initiative to ban H-bomb tests. It was passed unaminously, and was sent to the Prime Minister. Similarly, the district committee of the AEU handed the Mayor of Rotherham a resolution on civil defence which declared that 'leading scientists of high repute have declared that there is no civil or other kind of defence against such weapons.'[20]

But alarm was by no means confined to one end of the political spectrum. Cherwell received a request from Peter Emery, Conservative candidate for Lincoln in the General Election of 1955, to speak on the nuclear issue. The reason was that 'Mr Geoffrey de Freitas, Socialist Member of Parliament who has a majority of only 3,500, is using as the main plank of his campaign the Hydrogen bomb, as a means to try and scare people and thus vote Labour. I am doing everything to combat this but there is no doubt that the people of Lincoln are genuinely worried and concerned on the problem of nuclear energy.'[21] Cherwell was not in a position to speak publicly, but it is clear that he too believed public opinion was very disturbed after *Lucky Dragon*. He wrote to Churchill 12th May 1954 that:

> The 'ban-the-bomb' meetings and so on seem to be to show that the Socialists are trying to promote the view that their party alone is anxious to prevent H-bombs being used against this country. (Indeed I am told, such is the guillibity of the electorate that a great many votes in Edinburgh were cast on this basis.)[22]

These local protests were accompanied by restiveness in the Parliamentary Labour Party. On 5 April, the Labour Party forced a debate in the House of Commons, chiefly notable for Churchill's attack on Attlee for failing to secure US co-operation in atomic information. In spite of the indignation expressed on Labour benches at Churchill's jibe, it fairly summed up the real issues of nuclear armament as far as the Labour leadership was concerned. It was Attlee who had taken the lead in removing the question from public debate. He had treated public opinion lightly and his main preoccupation and, as Churchill said, failure, was on the issue of Anglo-American co-operation. The Labour Party was not, therefore, in a position to lead or direct the protests and opposition remained a grass-roots phenomenon with some parliamentary support.

The effect of the *Lucky Dragon* incident were also felt elsewhere. It led to the foundation of the Campaign Against Nuclear Tests, a forerunner of CND. The membership of this illustrates how 'middle opinion', or sections of it, were becoming more critical of government nuclear policy. This again can be illustrated in the case of Rotherham where on 28th June the United Nations Association proclaimed an anti-H-bomb week when, with the Mayor's support, petitions for a test ban were collected.[23] Similar moves were made among that relatively neglected, but important part of the British Peace Movement, the Non-Conformist Churches.[24]

The *Lucky Dragon* incident served to revive the scientists' campaign. On 25th April 1954, the AscW issued a statement by its executive committee. This proclaimed that the AScW

> Welcomes the widespread reaction throughout the world against the recent hydrogen bomb tests. It also notes with

misgiving a tendency to place the blame for these developments on scientists. We believe, on the contrary, the major responsibility lies firmly on the shoulders of those who hold political power. Nevertheless since scientists should have a greater understanding than others of the likely consequences of their work, they have a special responsibility to give a clear warning to the public on the implications of recent developments.[25]

The editorial went on to ask for the revival of disarmament proposals. It argued for the feasibility of inspection and, in particular, it asked Britain to take the lead in banning tests and in a diplomatic initiative to bring about nuclear disarmament.

Other groups of scientists also experienced a quickening of interest. The MAPW began to turn more attention in 1954 towards nuclear weapons. In that year it produced a pamphlet on the medical effects of atomic explosions. In 1958, the MAPW sold 2,000 copies of this to CND. The MAPW also wrote to the Foreign Office in 1955 warning of the medical dangers of radiation. By 1959, however, the MAPW was declining in influence. It was hit by a proscription by the Labour Party in 1958 — a very heavy handed way of dealing with an association of this type. Much more significantly, the members of the MAPW were drifting into CND.[26]

The effects of the *Lucky Dragon* was not simply to engage public attention but also to accelerate investigation of the genetic effects of fall-out. Governments remained very sensitive about this issue. In 1955 H.J. Muller was prevented by the American government from giving a paper on the effects of radiation on genes at a conference on Atoms for Peace at Geneva.

However, there was response to public disquiet. In 1955 the Genetical Society, which included L.S. Penrose, set up a panel of investigation, the first meeting of which took place at the end of April. By the end of 1955, however, the Society was superseded by a committee set up by the Medical Research Council (MRC). Against the opposition of some members of the Genetical Society, the Society suspended its own deliberations and sent its material on genetics and radiation to the MRC. In the same year two other investigations were set on foot, one by the United States National Academy of Sciences. Their report on the Biological Effects of Atomic Radiation was published in 1956. UNESCO also set up a committee on this question which reported in 1958. The MRC report appeared in 1956 and was followed by a further one in 1960.

The question of radiation was a highly technical one made worse by the fact that new information was becoming available contemporaneously to the committees, by changes in nomenclature and by the fact that the medical effects of radiation differed accordingly to which sector of the population was exposed to it. Thus by the mid-50s, it was known that children were particularly vulnerable to the accumulation of Strontium 90, a by-product of nuclear explosions, in their bones whereas the adult tolerance level would be higher. Moreover, just as there were multiple radioactive products contained in fall-out so there were multiple medical effects, not all of which were well known. In addition, the importance of the chain effect of fall-out, that is contamination of the environment eventually leading to further human exposure (i.e. the food chain), was only, slowly, being realised.

The interest of scientists in the 1940s and 1950s focused on defining the acceptable radiation dose for adults in industry. This had fallen quite dramatically over a number of years. In 1934, it was 1 roentgen (1r) per week (4r were enough to kill). In 1950, the International Committee on Radiological Protection (ICRP) lowered this to 0.3r per week. In 1959, it recommended 0.1r per week although it allowed for more intensive doses provided exposure was not continuous. Thus the ICRP recommendation of an allowable dose of 3r–10r for one week *only* was used by the British government as guidance for personnel during the bomb tests in Australia. The second major question in the 1950s was whether a 'threshold' existed below which radiation was harmless.[27]

The conclusion arrived at by the four investigations arising out of the fall-out controversy in the 1950s were, largely, the same. First, it was generally agreed that the sources of deleterious radiation were not bomb tests but natural background and medical radiation. The practical effect of the MRC report was to tighten up radiographical procedures in hospitals and elsewhere. Whilst radiation in general had increased, the MRC report of 1956 estimated that nuclear tests were responsible for only one per cent of total radiation. In 1960 it raised this to 1.2 per cent. The US Academy of Sciences calculated that the tests had produced an additional exposure of 0.3r for each adult over a lifetime. However, neither report accepted the threshold argument. Small compared with the effect of radiation from other sources, there were nonetheless appreciable and calculable effects in terms of genetic damage caused by tests.[28]

The problem was, as H.J. Muller put it, whether the appreciable effects were large enough to counter the perceived benefits of a nuclear arms policy. He, for one, believed that all radiation was deleterious. But he also believed that the USA's defence commitments were also important and, in his opinion, outweighed the bad effects through testing. But he admitted that this was a moral not a scientific judgement.[29]

The controversy over fall-out revealed a number of things. First, as far as Britain was concerned, there was ignorance about the effects of radiation and the logistics of fall-out. The government was dependent upon the limited supply of genetical knowledge and, to stir it into action on this question, on the existence of independent-minded and sometimes exasperated critics from outside government circles. Richard Doll, for example, made an impression at Harwell's conference on radiation in 1955 by his persistent attacks on the notion of a threshold. According to the Harwell committee on genetics and radiation his paper 'is in our opinion the most important because it is the only one bearing directly on the question whether the increased dose from nuclear explosions, however small, can be expected to have deleterious effects.'[30] Doll believed the onset of leukaemia was unaffected by length of exposure or timing. Whatever, differences of opinion existed, however, by 1960 it was admitted that fall-out had bad effects. As the geneticist Charlotte Auerbach pointed out, commenting adversely on Prime Minister Harold Macmillan's statement to the House of Commons in April 1959 about radiation, the objective should be to reduce as far as possible all risks from man-made sources of radiation, not to be complacent about it.[32]

In fact by 1960, the Medical Research Council had produced its second report which gave a much fuller picture of the degree to which dangers from radiation were increasing. The death rate from leukaemia was estimated as rising from 17 per million persons in 1931 to 42 per million in 1950 and 56 per million in 1959. This was thought to be largely due to medical radiography but it showed that tolerance of increased radiation was not unlimited.[32] Public awareness of this was increased by the accident at Windscale nuclear power station in 1957. Following a leak of radioactivity thousands of gallons of milk in an area of 100 square miles had to be destroyed because of radioactive contamination.

Significant developments towards internationalisation of the problem of nuclear weapons were being made by the WFSW. Since 1951, the executive and, in particular, its president Joliot-Curie, had been keen on re-establishing some form of international scientific co-operation to press for nuclear disarmament through the medium of a conference. In 1953 there had been unsuccessful approaches to Van Kleffans, secretary general of the UN.

In Paris, in January 1954, a few months before the *Lucky Dragon* incident, Bernal had met Eugene Rabinowitch, secretary of the American Association of Atomic Scientists and editor of the *Bulletin of the Atomic Scientists* to discuss the possibility of a conference. Rabinowitch was keen on the idea but he believed that it should comprise only Western scientists and of these, only the most eminent. When the WFSW met subsequently in Vienna in September 1954, the issue was discussed further. The Soviet scientist Oparin believed that a conference should discuss three things, the peaceful uses of nuclear power, control and inspection of nuclear weapons and the medical effects of nuclear weapons. Bernal, however, thought that, at this stage, a conference including US participation was almost impossible but he suggested tentative soundings should continue to be made but discreetly and without premature publicity.[33]

Nonetheless some movement took place.[34] Biquard and Joliot-Curie of the WFSW began sounding out scientific opinion. In Britain, Eric Burhop at University College, London assessed the situation for Biquard. The executive vice-president of the ASA, Joseph Rotblat was in favour of an international conference as were two other members, Kathleen Lonsdale and Harrie Massey. At this time Rotblat was encouraging the setting up of study groups of nuclear questions in the ASA. But, Burhop cautioned:

> I gather in the prevailing atmosphere it is difficult enough to get scientists to discuss these matters in groups among themselves and that it would be more difficult than ever to get them to come to a conference in which the WFSW played a leading part. Some leading American scientists such as Bethe are said to be against the whole idea of a conference, however organised, and I gathered, among influential British scientists, Mott was only lukewarm. The point of view was put to me that under present circumstances it would be easier to get Soviet scientists to attend a conference in the organisation of which the WFSW took no part, than to get American scientists to attend a conference partly organised by WFSW. Of the people with whom I have discussed the conference Rotblat . . . is, I think, keenest on the ultimate idea of a

conference and most disposed to associate with the WFSW but unfortunately he is not so influential as some of the others.[35]

The progress of this conference was hindered by the reputation of the World Federation of Scientific Workers (WFSW) as too left-wing. Nonetheless, early in 1955 there was a break-through. At Christmas 1954, Bertrand Russell made a broadcast signalling a considerable change of heart on the question of nuclear weapons. In his opinion they were now a formidable threat to humanity. Joliot-Curie approached Russell after this broadcast, but Russell was reluctant to join in the call for an international conference without the support of scientists who were eminent in their field and whose politics, as far as their respective governments were concerned, were above reproach. To gather these names together Joliot-Curie worked through his contacts in various parts of the world. These included Burhop, Powell and Rotblat in Britain, Otto Hahn and Max Born in Germany, and Eugene Rabinowitch in the United States.[36] After subsequent meetings with Joliot-Curie, a petition containing 60 eminent scientific names, including Einstein and Russell, expressing disquiet at the nuclear arms race and calling for international action was published (posthumously in Einstein's case). The Russell-Einstein statement of 9th July 1955, as it became known, signalled the increasingly 'respectable' character of the nuclear protest movement and helped to assuage the fears and reservations among some scientists about publicly expressing opposition to the nuclear arms race.

Two other events hastened the process. One was the Lindau Appeal. This appeal against the misuse of atomic energy was signed by 16 Nobel prizewinners and issued at Lindau, West Germany on 15th July 1954. In addition the conference of parliamentarians for World Government held in London in 1955 and attended by representatives from both the USSR and USA helped to spread the idea of the value of an international conference on atomic weapons. It also initiated contacts later to prove of value in assembling representatives from various countries.

The international conference, initially scheduled for India in 1956 (a project abandoned because of financial problems) was eventually held in the small Canadian town of Pugwash in 1957. This was due to the American millionaire, Cyrus Eaton, whose sponsorship solved the problem of financial viability. What was Pugwash meant to achieve? Basically its participants were chosen for their eminence, though with some misgivings (Burhop, for example, envisaged Pugwash as more of a mass movement of scientists); secondly, they were given a comprehensive agenda for discussion ranging from nuclear disarmament to the peaceful application of atomic power. Pugwash did several things. It helped to re-create the international scientific unity which had been severely disrupted since 1945. It also showed that the 'notables' of science were prepared to participate in this process.

To illustrate the magnitude of this achievement one could point to the fact that sponsors of Pugwash included Otto Hahn, who in 1953 had been prominent in the Hamburg Congress organised by the Congress for Cultural Freedom, an organisation sponsored by the CIA. Out of this conference the Committee on Science and Freedom emerged under the auspices of the Congress for Cultural Freedom. Its sponsors included Walter Gerlach and Max von Laue who, however, on 13th April 1957, were signatories to a protest by German scientists against the arming of the Bundeswehr with nuclear weapons.[37] Although the Committee on Science and Freedom was open to a wide range of political opinion (Bernard Russell was amongst its original sponsors) its objective was to counter the Soviet threat to scientific freedom. The tone of its leading participants and organisers reflected this more than the mood of its average participant, but it would not in 1953 have led or organised protests against the American test programme or German atomic armament. Subsequent events show that some of the scientists involved in the Committee were now prepared to be openly critical on both issues and to participate in protests side by side with left-wing scientists, most of whom would have found the tone and concerns of the Committee on Science and Freedom unsympathetic. This amounted to a preparedness on the part of middle-of-the-road scientists to be openly and actively critical, something which before 1954 had been unusual.

This shift in scientific opinion was significant for more than the scientific community. Pugwash heralded the kinds of co-operation required in the negotiations between scientists of both sides leading to the Test Ban Treaty of 1963. Moreover, Pugwash had other important functions. It showed the preparedness of scientists to educate and inform the public on nuclear issues. Until this development, apart from the statements made by small groups of politically active scientists, the information and knowledge disseminated to the general public about nuclear matters had been, particularly in Britain, of an appallingly low standard.

Increasing public concern about the effects of nuclear testing was heightened by Britain's growing commitment to a nuclear strategy. This policy direction was clearly and unambiguously stated in the Defence White Paper of 1957. So was Britain's extreme vulnerability to nuclear attack and the impossibility of providing an effective protection for her population in the event of nuclear war. In 1955, Britain's decision to proceed with the development of a hydrogen bomb was announced.[38] These events helped to accelerate opposition to nuclear armament; in particular, it revived opposition in the trades union movement. In July 1954, *The Scientific Worker* (the organ of the AScW) listed the trades unions in favour of banning the H-bomb. These included the Union of Shop and Distributive and Allied Workers, the Amalgamated Engineering Union, the National Union of Railwaymen, ASLEF the train drivers union, the Electrical Trades Union, and the Union of Post Office Workers. Also included in the list opposing British nuclear armament were a number of smaller unions, the Co-operative Party, and the Scottish TUC.[39] At the 1955 Trades Union Congress, these unions secured a motion calling for a government initiative to bring about nuclear disarmament.[40] This was forwarded to the Foreign Secretary but the General Council, when reporting back to Congress in 1956 on the progress of the resolution, indicated that they had a number of reservations about it. 'They were aware that in some respects it did not deal as adequately as some declarations of previous Congresses with certain aspects of the problem, for example, the need for parallel reductions in nuclear and conventional armaments, and the problems of an effective international control system.'[41] In other words the TUC Council reiterated its commitment to the United States position on Baruch.

Suspicion arose on the floor of Congress that the policy agreed in 1955 was not being rigorously pursued. This suspicion was justified. There existed a peculiar dialectic in the Labour movement between anti-nuclear rhetoric and the actual formation of policy. This is illustrated by the case of R.H.S. Crossman, one of the Labour Party's spokesmen on defence and in the 1960 Party Congress a 'compromiser', according to A.J. Groom, between the unilateralists and anti-unilateralists.[42] In the aftermath of the Civil Defence revolt in Coventry in 1954, Richard Crossmann wrote an article in the *New Statesman* criticising the government's Civil Defence policy — a rather broad target.

> Successive governments have stubbornly refused to face the fact that, once the American monopoly of the A-bomb had been broken, this Defence Policy exposed our civilian population to certain and instantaneous reprisals in the event of a general war.[43]

But Crossman's proposals were, in view of the rhetoric, surprising. He proposed 'passive defence', a phrase later used by Edward Teller, which implied that a nation, properly organised in shelters, could survive a nuclear attack — a proposition rejected by those who criticised civil defence policy. His criticism of the government was based solely on the grounds of its inadequate expenditure. In fact, in the Defence White Paper of 1955 the Conservative government subsequently increased expenditure on civil defence from £29 million to £70 million. Secondly, Crossman argued for the retention of the H-bomb as an essential part of British policy. What Crossman demanded was cutting expenditure in other areas of defence and less reliance on conventional forces. Thus Crossman called for a joint strategy: atomic forces were to cover Europe and the Middle East, and conventional forces, Asia and Africa. This view, apart from the notion of a demilitarised zone in Germany — a concession to the strong body of Labour opinion opposed to German rearmament — roughly anticipated the course of Conservative nuclear policy. Thus, whilst the rhetoric was anti-government, the reality was that the opinion formers in the labour Party held similar views on Britain's nuclear role to the Conservatives.

At the same time the growing volume of protest in the Labour Party about nuclear weapons eventually forced the leadership to re-evaluate its position. This is illustrated by the series of policy documents issued by the executive from 1955 onwards. These, whilst not conceding unilateralism, tried to head off criticism by, for example, calling for a test ban and criticising the government for failing to make a major initiative on disarmament. Nonetheless, there was a disingenuous character to these efforts. According to the Labour leader Hugh Gaitskell, writing in his diary in 1956,

> We have been in a jam on the H-bomb tests. We have supported producing the bomb and, obviously, if you want to produce it, you must be able to test it. At the same time, we

demanded the abolition of all tests. This was the position at the last election and still is supposed to be the party line. But I have managed to get the front bench away from that by getting them to ask for control rather than abolition of tests. I knew that sooner or later somebody was going to get worked up about this and ask why we were not demanding abolition, and, sure enough, this happened at one of the evening party meetings a little while ago. We then had a discussion in the Parliamentary Committee of a rather acrimonious kind, at which I had the utmost difficulty in making them see how inconsistent the position of the party was. It was clearly a subject which we might have to face up to, but which the leadership ought to avoid raising in Parliament because of our vulnerable position.[44]

In fact, Hugh Gaitskell was forced to admit in a House of Commons debate in 1957 on the Defence White Paper that he would not stop British nuclear tests unilaterally. This led to further unrest and a patched up compromise in which the Labour Party pledged itself to temporary suspension of British tests and an appeal to the USA and USSR to reciprocate. Later Gaitskell wrote of the controversy over Labour's nuclear arms policy in these years: 'I blame myself for not understanding sooner how strong the feeling in the party was.'[45]

By the middle 50s, from 1954 to 1957 approximately, the government the Labour opposition, and the TUC General Council were simply 'holding the line' against an increasing volume of protest against nuclear weapons. In one sense, the aim of the scientists involved in the opposition to nuclear weapons in 1945 had been achieved — a significant growth in public protests about the nuclear arms race. Nor was their contribution insignificant.

Britain's relationship to nuclear arms had changed significantly throughout this period. In the late 1940s, conventional rearmament was still the major preoccupation. Britain, none the less, aimed to develop an atomic bomb and hoped this would confirm its status as, at least, a second ranking power. The acquisition of a bomb by the USSR in 1949, three years before Britain, was a blow. Britain persevered and, although in the early 1950s it had considerable problems in building up a strategic bomber force with the capacity to deliver nuclear bombs to the territory of the USSR, the country was still optimistic about emerging as a major nuclear power. However, on the eve of Britain's explosion of an atomic bomb in

1952, the nuclear technology of both the USSR and the USA moved into the hydrogen phase. Thus Britain, after the explosion of its atomic bomb, was still, in spite of facile optimism in some quarters, in the position of having to catch up.[46]

By 1956 some of the problems seemed to have been overcome. Britain could look forward with confidence to having both a hydrogen bomb and an effective means of delivery by long-range bombers. The British Defence White Paper of 1957 was a recognition of this fact. It committed Britain to a mainly nuclear defence and looked towards the eventual reduction of conventional forces. There was talk, too, of intermediate ballistic missile systems. Politically this was an inept moment to announce this, given the gathering opposition to nuclear weapons, but militarily it could be argued that it was feasible. The Soviet Sputnik, on 4th October 1957, killed this strategy. It demonstrated that the Russians were well ahead of Britain in the development of missiles capable of delivering atomic warheads long distances, and nearer the USA in weapons development than the most pessimistic military planner had supposed. As far as Britain was concerned, the USSR had superior facilities for a policy of nuclear deterrence. Moreover, this third successive phase of technological development was likely to be an expensive and difficult one for Britain. This spurt in nuclear technology more than ever reaffirmed Britain's relative decline in status since 1945. Efforts were made to build an independent British missile, but they were unsuccessful. And what guarantee could there be that Britain's effort would not be trumped again as the nuclear arms race escalated yet again into another technological phase? Thereafter, Britain's nuclear policy, as the White Paper of February 1958 admitted, depended upon admission to the nuclear secrets of the USA. The nuclear role of Britain would henceforth, certainly as far as delivery of nuclear warheads was concerned, be a subordinate one, or as the 1958 paper put it, an 'integrated' one:

> Within this integrated structure, it should be possible gradually to get away from the idea that each member nation must continue to maintain self-contained national forces which by themselves are fully balanced.[47]

By 1958, the climate in which the peace movement operated had substantialy altered. In 1957, the events leading to the formation of the Campaign for Nuclear Disarmament had taken place. There had been a number of unsuccessful forerunners to CND. In the aftermath of the *Lucky Dragon* incident, local committees for the Abolition of Nuclear Weapons had been formed comprising, among others, trades union officials, women's co-operative guilds and the Society of Friends. These wrote to the Foreign Office in November 1955 warning of the dangers of nuclear radiation. But the movement fizzled out. By 1957, however, several things had changed. A real and vocal opposition had emerged within the Labour Party against the party's strategy on nuclear weapons and especially the failure to give unequivocal support to a test ban. Resentment was also building up even more explosively at the TUC. So long as the 'old guard' on the General Council held sway, the anti-nuclear protests at the TUC were given short shrift. For example, in 1957 the General Council responded to pressure from Congress by repeating what the United States government position at the recent disarmament talks at Geneva had been. The International Confederation of Free Trades Unions (ICFTU) outlined this in the following way:

> To be really effective, disarmament agreements should be paralleled by settlements of the most urgent problems affecting the peace of the world, such as the reunification of Germany . . . and the establishment of genuine peace in the Middle East.

The General Council's line echoed this: 'It was unlikely', it argued, 'that it would be possible to proceed far along the road of general disarmament without settlement or compromise in certain danger areas of the world.'[48] But when Frank Cousins, a unilateralist, replaced Deakin as head of the largest union, the Transport and General Workers Union (TGWU), the balance of power in the TUC shifted dramatically.

This building up of frustration reflected something that the 'old guard' and the Labour Party executive failed to appreciate, that from the *Lucky Dragon* onwards, real apprehension existed over the effects of nuclear weapons and especially their testing. Further, the nuclear arms race which got under way in the 1950s seemed to many to indicate a dangerous failure of judgement on the part of Western governments — a view reinforced by the *Lucky Dragon* incident. Lastly, the cold war had created a way of dealing with political opposition which ultimately produced a reaction. If it was not respectable and acceptable to voice opposition to the government's policy on nuclear weapons, and if protest led to exclusion from influence and to accusations of disloyalty, then nothing was left but a grass-roots movement. By keeping away the scientists opposed to nuclear strategy from positions of influence, by concealing major facts about nuclear weapons, by the policy of the TUC of using the atomic bomb as a 'test' of loyalty and isolating its opponents, the system was generating a degree of discontent and frustration that would eventually find expression in a new form of organisation: the Campaign for Nuclear Disarmament.

Notes

1. See the reports in *Atomic Scientists Journal*, vol. 3, No. 5 (May 1954), pp. 292-7.
2. Ibid., pp. 296-7.
3. *Manchester Guardian*, 13 October 1953.
4. See J.H. Webb, 'Fogging of photographic film', *Physical Review*, vol. 76, no. 3 (1949).
5. H.J. Muller, 'Artificial transmutation of the gene', *Science*, vol. 46, nos. 84-87 (1927).
6. See Robert Milliken, *No conceivable injury: the story of Britain and Australia's atomic cover-up* (Penguin, Harmondsworth, 1986) chap. 7.
7. Merrit Eisenbud and John H. Harley, 'Radioactive dust from nuclear deteroriations', *Science*, vol. 117 (13 February 1953), p. 141.
8. W.G. Marley, 'Radioactivity and civil defence', *Atomic Scientists News* (May 1952), p. 193.
9. Milliken, *No conceivable injury*, fn. 6, pp. 124-9.
10. Marley, 'Radioactivity and civil defence', fn. 8.
11. Penrose Papers, 79, Penrose to Loutit (26 October 1950).
12. Ibid., Loutit to Penrose (26 October 1950).
13. Ibid., Loutit to Penrose (28 March 1950).
14. Milliken, *No conceivable injury*, Fn. 6.
15. Cherwell Papers, J146 (18 November 1954).
16. CAB 129/72, C(54) 389. Fall Out, Memo by the Minister of Defence (9 December 1954).
17. CAB 128/27 CC(54) 86, 53rd Conclusion, Minute 3 (14 December 1954).
18. *Coventry Evening Telegraph* (6 April 1954).
19. *South Yorkshire and Rotherham Advertiser* (24-30 April 1954); *Newcastle Journal*, (13-14 April 1954); *Manchester Co-Operative News* (3 April 1954). The Woman's Co-operative Guild sent a resolution to the Prime Minister and Foreign Secretary stating: 'We therefore demand the British government should take urgent action asking the Americans to cease further tests.' For reports of activities in other areas, see *Atomic Scientists Journal*, vol. 3, no. 6 (July 1954), p. 362.
20. *South Yorkshire and Rotherham Advertiser* (24 April 1954), p. 5.
21. Cherwell Papers, J156 (10 May 1955).
22. Ibid., Cherwell to Churchill (12 May 1954).
23. *South Yorkshire and Rotherham Advertiser* (3 July, 28 August 1954).

24. Report of the Sheffield Methodist May Synod, *Sheffield Telegraph* (8 May 1954). The links between religious non-conformity and political radicalism in certain areas of England were still strong in post-war Britain.
25. *The Scientific Worker*, vol. 9, no. 3 (May 1954), p. 24.
26. MAPW papers in Penrose Papers, 41/3 Memoranda.
27. Penrose Papers 79.
28. *The hazards to man of nuclear and allied Radiation*, Medical Research Council, Cmnd 9780 (HMSO, London, 1956). For discussion of the controversy over nuclear radiation in the USA, see Carolyn Kopp, 'The origins of the American scientific debate over fall-out hazards', *Social Studies of Science*, vol. 9 (1979), pp. 403–22. *The hazards to man of nuclear and allied radiations*, second report to the Medical Research Council, Cmnd 1225 (HMSO, London, 1960). Also reported in *British Medical Journal*, vol. 2 (December 1960), p. 1947 and Genetic effects of atomic radiation (National Academy of Sciences of America) *Science*; vol. 123 (29 June 1956).
29. H.J. Muller, 'Genetic damage produced by radiation', *Science*, vol. 121 (17 June 1955), pp. 837–40.
30. Penrose papers 79, MRC Committee on Medical Aspects of Radiation, MRC55/638.
31. Charlotte Auerbach, 'The Prime Minister seems to think it a cause for complacency that the human race has survived in spite of a considerable amount of radiation from natural sources. The opposite is true . . . The fact that there is this inescapable amount of natural radiation . . . makes it even more imperative to keep radiation from any additional sources as low as possible', 'Radioactive fall-out', *Nature*, vol. 183 (27 June 1959), pp. 1773–6.
32. *Hazards to man*, second report to the Medical Research Council; also reported in the *British Medical Journal*, vol. 2 (December 1960), p. 1947.
33. WFWS Papers, MSS270 Notes taken by Pierre Bicquard, Vienna (12 September 1954).
34. Simultaneously the United States and six other nations announced a plan for an international agency to encourage projects for the use of atomic power for peaceful purposes. This Atoms for Peace agency held a conference at Geneva in 1955. Although Gilpin says that the decision to set up this agency preceded the *Lucky Dragon* (*American Scientists and nuclear weapons policy*, Princeton University Press, New Jersey, 1962, chap. 5. fn. 5), this incident certainly accelerated it. The objective of Atoms for Peace was partly, to improve the public image of atomic energy. It did not deal with questions of atomic weapons. See *Nature*, vol. 174 (16 October 1954), p. 724, vol. 176 (29 October 1955), p. 814.
35. Burhop Papers, Burhop to Biquard (30 November 1954), Fn. 43.
36. 'Actions of the WFSW leading up to the first Pugwash Conference', E.H. Burhop (Burhop Papers (1960).
37. Otto Hahn, *My Life* (Macdonald, London, 1970), pp. 210–28.
38. Statement on Defence, Cmd 9391 (HMSO, London, 1955).
39. *The Scientific Worker*, vol. 9, no. 4 (July 1954), p. 4.
40. Motion 7, TUC Reports, Southport (1955).
41. TUC Reports, Brighton (1956), p. 195.
42. A.J. Groom, *British thinking about nuclear weapons* (F. Pinter, London, 1974), Chap 5, Fn. 5, p. 454.

43. Richard Crossman 'Coventry and the H-bomb', *New Statesman* (5 June 1954), p. 721. For Crossman's views on defence, see 'The dilemma of the H-bomb', *New Statesman*(26 February 1955), p. 268 (with George Wigg).

44. Philip M. Williams, *The diary of Hugh Gaitskell*, 1945-56 (Cape, London, 1983), p. 557.

45. Philip M. Williams, *Hugh Gaitskell* (Cape, London, 1979). p. 453.

46. For example, 'News of Britain's latest success at Woomera will increase the uneasiness of American scientists already afraid that British atomic research workers are outdistancing them,' *Daily Herald* (15 October 1953).

47. Report on Britain's Contribution to Peace and Security, Cmnd 363 (HMSO, London, 1958), p. 4.

48. TUC Reports, Blackpool (1957), p. 202-3.

7
The New Right

The proponents of virulent racism were increasingly isolated in the 1950s. This was not only due to the legacy of the war and the efforts of UNESCO. Other important political changes were taking place. In the United States the 1940s saw an increase in black political awareness manifesting itself in the movement for voter registration both North and South.[1] Other areas of black life were affected. There was a decline in the practice of segregation in some parts of the United States military. In addition the attempt to restore the idea of racial equality in intellectual life was bound to affect American blacks.

Unwittingly Ruggles Gates walked into the maelstrom caused by these events. In 1945, he left England where he had been Professor of Botany at Kings College, London, for a visiting fellowship at Harvard. In the spring semester of 1947, he was invited by H. Ellinger to give a series of lectures in the zoology department at Howard, the black university in Washington. Ruggles Gates saw this as an opportunity to collect evidence for a new book he was preparing on *Pedigrees of Negro families*. Before he had been at Howard long, a mass meeting of staff and students was convened to demand his resignation. 'Within 48 hours', wrote Ruggles Gates, 'I tried to address them and was nearly mobbed.'[2]

In February of 1947 following the meeting, the Dean of the College of Liberal Arts, Dr J. St Clair Price, received a letter from staff and students demanding Ruggles Gates' dismissal,

> As a person whose ideas and writings imperil the integrity and self respect of the Howard University community. We offer in evidence the following:
> Professor Gates rejects the notion that all races belong to

Homo Sapiens but believes that, 'The living "races" of man are not contemporaneous emergents from a common stock but are of different ages, the Australoids being the oldest, then the Bushmen, followed probably by Negroids, Mongoloids, Caucasoids in that order.'[3]

The letter went on to quote from other articles of Ruggles Gates in the 1930s to illustrate his views on the inferiority of the black race.

Ruggles Gates left after a term but the affair rumbled on. Shortly after, Ellinger was forced to leave because of continuing unrest. He vented his feelings in a series of bitter letters to Ruggles Gates. The president of Howard wrote unsuccessfully to Ruggles Gates to demand the return of the Negro pedigrees which Gates had collected at Howard. Ruggles Gates began to experience a feeling of increasing isolation compounded by a sense of persecution due to events at Howard, which he attributed to a Jewish conspiracy.

> I had been in Washington a week when I discovered that the Jews were in effect, controlling the university for their own purposes. They soon attacked me, dug up and circulated excerpts from a paper on populations which I published in London 14 years ago, in which I had indicated that the Negroes as a race were not on a par with whites — On this base the students held a mass meeting in which they demanded my dismissal.[4]

Gates' problems were increased by the tardiness of Blakiston Press, the publisher of *Negro pedigrees*. Alarmed by Howard University's attempts to retrieve the pedigrees and by fear of a nasty controversy, they tried to get Ruggles Gates to 'insert a statement in the front to the effect that all the evidence herein tends to show that there is no difference between the Negro and any other race in regard to qualitative or quantitative characteristics to which the rest of mankind is subject.'[5]

Gates reacted strongly against this. He also suspected that he was being shut out of the academic press in general. He was vehement about the reviews of his work — which were generally couched in more hostile tones than had been the case pre-war — and peppered editors with complaints. The results were not good. In 1949, the editor of *Nature* L. Brimble, wrote back in reply to his complaints, 'Your letter of April 1st was most offensive. The accusations it contains were so preposterous that I shall not bother to deny them. I would, however, have liked the opportunity to publish your views on the Jews in the United States.'[6]

In the atmosphere of the late 1940s, Gates' defence of his views was disastrous. But his isolation was shared by others. In 1948, Henry E. Garrett, a psychologist at Columbia and a proponent of Negro inferiority in intelligence, wrote to Ruggles Gates asking him to defend him in *The Scientific Monthly* against attacks on his ideas. Gates replied rather brusquely[7] but, none the less, in the late 1940s and 1950s a group of psychologists and anthropologists whose outright racism had alienated or offended many of their colleagues began to draw together, with the ultimate aim of relaunching their attack on egalitarianism.

Although the 1950s were a period of isolation for the proponents of racial difference, many were on the look-out for causes to which they could attach themselves. In the United States of America, the Supreme Court Decision in 1954 declaring segregation illegal in Southern schools provoked them into action. In the South, this decision led to the setting up of citizens defence groups among whites opposed to desegregation. Ruggles Gates, still at Harvard, wrote to one of these — the Jackson Citizens Council (Mississippi) sympathising with their plight. They wrote back,

> As a result of the pseudo-scientific brainwashing which has been used so copiously in our colleges and theological seminaries for the past decade or so there are many white people, even in the South, who would not oppose the disastrous aims of the Communist and Socialist controlled negroes . . . I am taking the liberty of placing your name on our mailing roster if you have no objections, so that you will be kept informed of the nature and purposes of this organisation. You may be interested to know that we are in correspondence with a number of outstanding scientific and literary men who are beginning to be seriously concerned about the ultimate consequences of the present racial drift.[8]

Ruggles Gates was reluctant to be publicly associated with the Jackson Citizens Council but he was very willing to prime them with appropriate books and articles. In appreciation, the Council's secretary W.J. Simmons wrote,

> You were kind enough to recommend Bryan Campbell's

book, *American race theorists* which has been no end of help to us. May I return the favour by recommending, if you have not already seen it, *The diminished mind: a study in planned mediocrity* by Mortimer Smith. I expect you have seen the new book by Sorokin *Fads and foibles in modern sociology*. His book does much to blast the ground completely away from the entire basis of the Supreme Court decision.[9]

Subsequently, Ruggles Gates frequently discussed at length the evils perpetuated on the South by the North. Apart from this the more permanent legacy from these contacts in the 1950s was a network of organisations and individuals brought into contact with each other. These, often generously funded, were dedicated to pushing back the tide of racial egalitarianism in intellectual life. They were composed of a combination of types: disaffected Southerners and their intellectual representatives and anti-Communists who saw racial equality as a concession to Communist egalitarianism. As the 1950s progressed, the American representatives of these currents of ideas made contact with Britons unhappy at the West Indian (and later Asian) immigration into the United Kingdom and other Europeans prominent in pre-war eugenics, some of whom were compromised by their association with it.

The institutions which sprang up in the 1950s and 1960s in opposition to the greater egalitarian climate in the post-war world were numerous and they had a strong international flavour. In the 1950s the Institut International de Sociologie was founded. This according to James A. Gregor, a prominent member, was an organisation which 'has become increasingly aware of the importance of concepts pertaining to race for its sociological theory building.'[10] Other organisations included the International Association for the Advancement of Ethnology and Eugenics (IAAEE) whose members included Gregor, Corrado Gini (who had been important in Italy's pre-war eugenics movement), Henry Garrett (a psychologist at Columbia University) and Donald Swann. These institutions multiplied, often with interchangeable membership. They included the Foundation for Human Understanding, the Institute for the Study of Education and Differences, and the Institute for the Study of Man directed by Donald Swann. All had 'respectable' academic titles and aped the academic and research worlds. For publication outlets, they had at their disposal a number of small print houses, in particular the Noontide Press.

The new bodies also had funding. In particular the Pioneer

Fund, founded in 1937 by Frederick Osborn, secretary of the American Eugenics Society and Harry Loughlin, director of the Eugenics Record Office, gave grants for research in the area of human and racial differences. In the 1950s and 1960s, the Pioneer Fund was directed by Francis E. Walter who was chairman of the US House of Representatives Committee on Un-American Activities. Two other members of the Pioneers were Henry Garrett and James O. Eastland, a senator from Mississippi. Money was also available from Wycliffe Draper, a textile manufacturer. His death in 1972 released $1.4 million to the Fund. Whilst alive he funded several ventures from the profits from his textile fortune.[11]

The Draper connection was important in the foundation of the periodical *Mankind Quarterly*, which eventually appeared in 1960. Wycliffe Draper supported a Human Genetics Fund of New York and Henry Garrett wished to use this to set up a periodical which he hoped would allow the presentation of his views within a more scholarly apparatus. Garrett wrote to Ruggles Gates about this project,

> I was able to get the Human Genetics fund of New York to agree to make a grant which would cover publication costs for at least the first year . . . I think it is highly desirable that we may overcome the idea which many Americans have that all those who believe in race differences burn crosses and go around in bedsheets. Our friends Klineberg and Montagu *et al* have done their bit imparting this view.[12]

Eventually $2,500 were made available and R.R. Gayre in Edinburgh became the editor of the new periodical. Gayre was of the right but in a distinctly European tradition. He was pro-Monarchist and pro-Catholic (he was appalled by the Ulster Protestants, whom he considered developed typically into the worst kind of religious sceptic). His politics were based on the rejection of the French Revolution and its consequences. His contacts with the Italian and European right were also considerable. During active service in Italy during the Second World War, he had been responsible for the installation of Professor Martino as rector of Messina University.[13] Professor Martino became a prominent patron of *Mankind Quarterly*. Archaic as many of his beliefs seemed, or would seem to the American right, Gayre took a strongly pro-scientific stance, at least as far as race was concerned. Gayre suggested that the function of *Mankind Quarterly* should be to lay the scientific argument before the public:

> Several people there may be anxious to have published some short articles on various problems devoted to showing that race does exist; and that racial characters, physical, emotional and mental are transmitted. I think these are the points we have to establish in the minds of all intelligent educated people. They were once believed; but the constant propaganda of recent years has eroded them away, and our task will be to re-establish the truth of these facts in the minds of everyone. It is the battle of Darwin *vs* Lamarck and we have to show that Darwin was right and not the glosses on Darwin which are now being put around in the various re-interpretations of him.[14]

The editor did not feel it prudent to open the pages of the new journal at this stage to all who held this view, feeling that the respectability of *Mankind Quarterly* would be damaged if this were done. With this in mind he corresponded with some of his European contacts:

> I had a long letter from Professor Hans Gunther whom I knew before the war, giving me a complete list of names of people in Germany and surrounding countries, who are either afraid to write, because of political persecution, perhaps in some cases because they compromised themselves with the Nazis and others who are prepared to write. I have written back to him to explain our policy with regard to Germans, but I have said that bit by bit, where articles can be given to us which have not a political application, we shall be able to consider them and bring them in due course.[15]

Mankind Quarterly was intended to be a forerunner of other enterprises, including school text books and academic books. A symposium planned by its editorial board, entitled Outline of Race, was to include R. Gates, C.D. Darlington, Eugen Fischer and E. Raymond Hill speaking on race and biology. On race and psychology, contributions were planned from Fritz Lenz, E. von Eickstedt, Walter Scheidt, Corrado Gini and R.A. Fisher.

The appearance of *Mankind Quarterly* however, produced controversy. *Man* condemned it and it also produced a furious reaction from Juan Comas, the Mexican anthropologist, in the pages of *Current Anthropology*. *Mankind Quarterly*'s objectives were clear. They wished to discredit the UNESCO statement of 1950s and they were keen to exploit the differences about it which had appeared amongst academics when it was drafted. Whilst their aim of 'respectabilising' race doctrine was clearly unlikely to succeed in the early 1960s, they claimed a modest success in putting the question back on the agenda. James A. Gregor considered that, 'The fact that legitimate controversy has, once again, become possible at all, in these areas, augurs well for the science of society. That instances of a lack of academic detachment and academic etiquette were notable in their intensity if not in their abundance, does not alter the fact that significant progress was made towards restoring the lost equilibrium of serious social enquiry.'[16]

Why should *Mankind Quarterly* and its sponsors fight so fiercely over their right to claim their racial beliefs as part of science? This is especially significant since the appeal of, particularly the European right, was clearly to dimensions of human experience other than that of science. They claimed to value emotion and intuition as opposed to reason and investigation. The French right, in particular, attacked materialism and some on the right wanted an increase in religious authority in society. Why then fight over the question of whether Darwin or Lamarck were right? Why not settle questions of politics without reference to scientific authority?

One major reason was the power of science — especially the human sciences — as a means of distributing social roles, rewards and punishments, or at least of justifying the way they were distributed. In the 1960s this function of science came under attack and in the 1970s the right was to re-emerge as defenders of this function.

In the United States the attack on the human sciences came from several sources. One was the controversy arising from the use of the Minnesota psychological tests for personality evaluation (the MMPI). These were used by the State Department in the early 60s for candidates to the Peace Corps set up by President Kennedy. Senator Sam Ervin (North Carolina) launched an attack on the use of tests by government in 1964 during the debate on the Civil Rights bill. This was followed by a Congressional Inquiry into their use by the Civil Service Commission and the Department of Labour for the selection of federal employees.

The apprehension caused by the tests in the Peace Corps was

not solely on the grounds of their validity, although this was also disputed fiercely. One psychologist in evidence to the Congressional Committee considered that 'a test like the MMPI probably picks out people who either know how to or are willing to carry on superficial lying in a situation where it is more or less socially approved. The application of such a test to the Peace Corps means that you probably have in the Peace Corps a great number of people whose prime trait is the ability to lie in a minor way in situations where it is socially approved.'[17] However, the public outcry caused by the MMPI was chiefly on grounds of the invasion of privacy (there were questions on sex and family life) and on the dubious constitutionality of inquiries about religious attitudes. In addition, the tests represented a rather East Coast, Ivy League view of 'desirable' personality. For example some of its questions were intended to weed out the over zealously religious. For a country which was founded by religious enthusiasts and which still contained many religious fundamentalists, this seemed a rather narrow view of acceptable Americanism.

There was something else. Mainstream America often had a strong view of the importance of opportunity and 'getting on'. These tests as an agent for social pre-destination usurped rights which some Americans felt belonged to God alone. They certainly did not feel that the professional psychologist should usurp the role of Providence or stand in the way of self betterment. Much to their surprise therefore, the American Psychologists Association found itself on 4th June 1965 picketed by about twelve pickets from the Committee to Bring Morality to the Mental Professions.

For black Americans, attention focused on the role of intelligence tests to limit occupational opportunity. In his evidence to the Senate and Congressional Committees, Arthur H. Brayfield, president of the American Psychologists Association, stated his belief that psychological tests helped unrecognised talent and broke down religious, racial, sex and age barriers. 'I know of no other professional tool which has matched the effectiveness of individuals to realise their civil and human rights,' he testified[18] However the experience of black Americans was different and a series of court cases in this area highlighted this.

In the summer of 1963, a black American, Leon Myart, was refused a job at Motorola in Illinois. He claimed this was because he had failed an IQ test. The examiner who heard his case found that although Myart had failed the test, the test itself was not valid. This was because it 'does not lend itself to equal opportunity to qualify for the hitherto culturally deprived and disadvantaged groups.' Further it was obsolete in the light of today's knowledge 'because its norm was derived from standardisation on advantaged groups.'[19] On the review of the case by the Illinois Commission of Fair Employment (which claimed Myart had passed the test anyway) they did not demur from this judgement and Motorola was ordered to cease test Number 10.

Similar cases followed and influenced the framing of both the 1964 Civil Rights Act and the Equal Opportunities Act of 1972. Overall, the US federal government would not countenance the stopping of these tests. But it compromised with their critics. In 1970, the Equal Employment Opportunities Council stated that, to be acceptable, tests had to be properly validated and shown to be useful; suitable alternative procedures also had to be available. Tests were ruled unlawful if it could be shown that their intention was to discriminate. These ideas were incorporated into the new Equal Opportunities Act of 1972.

When intellectuals took stock of these developments they came to different conclusions about their relevance. The *American Psychologist* responded in the 1960s by becoming more self critical about its discipline and opening its pages to radical critiques of orthodox psychology. But in some respects the right in America could make common cause with the left on this issue. After all if psychological testing was the tool of 'progressive' capitalism — justifying the distribution of roles and rewards — then those marginal to American capitalism suffered too. This included those excluded from metropolitan culture, religious enthusiasts, smallholders, and small-town people as well as the urban disadvantaged and leftist critics of capitalism.

However, those groups who were 'marginalised' on the right by American capitalism and took refuge in hatred of metropolitan culture were never at the centre of the intellectual right. As H.E. Garrett said in 1959, the intellectual right's project went beyond 'burning crosses and white bed sheets'. Similarly, whatever sympathy was expressed by Ruggles Gates for the Southern cause in the 1950s, he took steps not to be too publicly associated with it. Their aim was to get into the centre and not the periphery of political debate. For the right, therefore, the use of science as a means of social determination of position — whether of class or race — was a powerful tool which they had no intention of surrendering. Moreover it gave them access into the institutional practices of government. It was these which they wished to put to work in their interest. They wanted to imbue the scientific spirit of modern capitalism with their ideas and this was as true of the sentimental critics of capitalism in France as it was elsewhere. But their opportunity to do so was severely limited by their political isolation. It was the gradual change in this which brought about the 'new right'.

In Britain, the eugenics movement had floundered in the 1940s, not merely through the revelations about Nazi eugenics, but also because of the arrival of full employment and welfarism. This had undermined the eugenic gospel. A new and more egalitarian mood swept Britain in the 1940s and it removed much of the *raison d'être* of eugenic social philosophy. This *raison d'être* was based on a hostility to welfare expenditure that raised working class levels of consumption. Eugenics offered an alternative means to deal with social problems primarily by raising individual fitness and rooting out the 'unfit'. The Eugenics Society secretary C.P. Blacker was well aware of these changes. In 1952 he published *Eugenics, Galton and after* which he saw as a means to rehabilitate eugenics. By concentrating on the nineteenth century liberalism of Francis Galton (with its emphasis on equality of opportunity, anti-aristocratic prejudice and anti-chauvinism) Blacker was able to claim that, 'This is an enlightened programme which has something in common with that which a socialist government is trying to put into effect today.'[20] Blacker admitted that social class had been prominent in the writings of several leading eugenicists in the years after Galton's death. He felt that the language they used 'gave offence to many social reformers'. However he could not bring himself to forgive critics of eugenics like Hogben, whom he described as leader of an 'anti-eugenic school of experimental biologist'; nor Haldane, to whom he referred only once and that in connection with Lysenko.

This revision of eugenics was not acceptable to all in the Society. In 1960, J.R. Baker complained to the new secretary C.G. Bertram. He considered the Society had become too indifferent about putting forward a bold eugenic policy, and had bowed too much to modern eugenic propaganda about the equality of all men of all races.[21] C.D. Darlington and Sir Charles Darwin, both long term members of the Eugenics Society, agreed. Both involved themselves with *Mankind Quarterly* when it appeared.

Baker was particularly exercised by the question of coloured immigration into Britain. In the 1950s, this was predominantly West Indian, but in the 1960s immigrants from Asia and Africa also began to arrive. Baker had begun collecting material for a book on race since the 1930s. In the 1960s, he put together his material, and sought a publisher. After a number of false starts and bitter disappointments, his book *Race*, dedicated to C.P. Blacker, was published in 1974 by Oxford University Press. *Race* became, along with other publications in the late 1960s and 1970s, one of a number of attacks on racial egalitarianism.[22]

Because of his interest in these questions, Baker found himself drawn into a network of, in some cases, very right-wing individuals.[23] Blacker, whose political sense was stronger, warned him against these. There were, he told Baker, dangers involved in associating with such people. The reception of Baker's book *Race* might be affected adversely. Subsequently Blacker warned Baker against involvements with Gayre and with the Noontide Press which Baker hoped might republish *Race* in the United States.

Baker did not heed his advice and was rapidly drawn into a circle of often eccentric American rightists. Baker's work brought him into contact with Carleton Putnam and, for a while Wilmot Robertson, both involved in orchestrating the American intellectual right. Whereas Carleton Putnam and Baker found many points of agreement and friendly relationships were established, relations with Wilmot Robertson became strained because of Robertson's antisemitism. In particular Baker found it incomprehensible that the views of his friend Professor Michael Polanyi could be attacked because of his race. He regarded Polanyi as an essential figure in the intellectual offensive against the left. Robertson, however, clearly regarded the old fashioned liberalism of Polanyi as unacceptable. He referred slightingly to Polanyi's career as Minister of Health in the Karolyi government which preceded the Hungarian revolution of 1919 led by Bela Kun. Similarly, Robertson expressed doubts whether Baker's other example of a Jewish friend of the right — Hans Eysenck — was in fact Jewish. Baker belonged to that section of the British right — inegalitarian, with strong views on race, advocate of a free market economy but rather puzzled by antisemitism. Communication with Wilmot Robertson broke down over this question.

Wilmot Robertson had more success however with the French right, for whom he acted as a conduit for ideas and new publications. In 1977 he was in correspondence with the French rightist Alain de Benoist and making trips to Paris to discuss French translations of American books on race and intelligence. Robertson was also responsible for spreading enthusiasm about sociobiology. He was, according to Putnam, very impressed by E.O. Wilson's book

Sociobiology. This enthusiasm was not shared by Baker or Putnam. Putnam, in particular, could not find the word 'race' in Wilson's index. In addition he had read in *Time* magazine Wilson's rejection of the label 'racist'. This he felt amounted to Wilson's self exclusion from their side of the political spectrum.

In spite of their misunderstandings and wilder extravagancies, the intellectual right began to reassert itself in the 1970s and 1980s. What was new about the 'new' right were the social and political circumstances which favoured its re-emergence. These circumstances differed from country to country but there were certain common intellectual currents in all countries. Of these the search for scientific authority for their political beliefs was vital. However the crucial intellectual dimensions of the 'new' right were to a large extent, shaped by the political conjuncture. This decided what was or was not intellectually acceptable.

The events of the 1960s had caused shock waves in many societies, the result of which was to produce a conservative reaction. In France, the end of the Algerian War in 1962 — which had divided the right — the 'Great Fear' of 1968 and the election of the conservative Giscard d'Estaing in 1974 all had their effect.[31] But other factors also played a part. The moral authority of the Resistance decreased as memories of the Second World War faded. One beneficiary of 'resistance' politics had been the Communist Party of France. But in the 1960s, it felt its hold weaken among intellectuals as the significance of the Resistance was either forgotten or consciously attacked. In many European countries, Britain for example, the rise of the new right coincided, less with the Cold War of the 1950s, than with the revival of the political divisions about the significance of the 1939–45 war. In Britain this took the form of criticising the political logic of Britain's involvement in a war against Germany and of criticising the social legacy of the 1940s, the welfare state and full employment. Both were said to have knocked Britain off course and precipitated her decline. With this came a re-evaluation of the 1930s as a period of modest but significant economic progress and a more favourable re-assessment of 'appeasement'.

In France, the strengthening of capitalist industrialisation in the 1960s was another factor. For many, 1968 was the last fling of radical anarchism. Some of the representatives of 1968, drawn from the anarchist tradition, later made peace with the right. For them, the radical right rejected proletarianisation and the bourgeois social order that accompanied it. Whereas the left, as represented by the

Communist Party, embraced industrialisation and large-scale bureaucracy. So a common theme emerged of the defence of the individual against the state and of the traditional patriarchal values of the French peasant family. The representatives of these opinions were destined to veer widely across the political spectrum.

In fact the French 'new right' was a peculiar amalgam of scientism and romantic emotionality. But then so was fascism itself. Two magazines were published, which could properly claim to be inspired by the new right. One of these was *Nouvelle Ecole* which appeared in 1971. This was published under the imprimateur of GRECE (Groupement de recherche et d'étude pour la civilisation européene). Alain de Benoist (a pen-name for Fabrice La Roche) was associated with this. The other was *Eléments*, a quarterly review published by GRECE. Both contained articles on biology and eugenics.

These esoteric magazines had a limited appeal. Eventually, however, ideas propounded in them were to find their way into much wider circulation newspapers and magazines. In the 1970s, the 'new right' found more fertile ground for the popularisation of their beliefs than they had done since 1945.

The connection between the American right and Benoist — through the medium of Wilmot Robertson — allowed Anglo-Saxon racial and eugenic thought to be channelled into French newspapers. For example, Alain de Benoist published an article in the French conservative newspaper *Figaro* on 30 June 1979 in which he discussed British and American sociobiology, including the work of E.O. Wilson, Richard Dawkins and Robert L. Trivers. But the British and American tradition was filtered through the traditions of French rightism. For example Pierre Vial, secretary general of GRECE, claimed in an article in *Le Monde* on 24th August 1979 that GRECE was not social Darwinist in a traditional way. This was because they rejected both reductionism in philosophy and also social hierarchy based on money. In contrast the views of someone like Konrad Lorenz were much more palatable. Lorenz had developed a biology with a clear political message. Modern urban civilisation had 'domesticated' the human race and, as in the case of the domestication of animals, this had led to physical, intellectual and spiritual degeneration. GRECE, if Vial is to be believed leaned towards a Nietzschean view of race, which recalls Baur, Lenz and Fischer's depiction of the *Übermensch* as a lonely warrior besieged by the money-grubbing cosmopolitan. This was far removed from the Anglo-American admiration of the successful businessman. But this view did not preclude extensive borrowing from British and American work on mental testing, eugenics and sociobiology. The Comité de Patronage of *Nouvelle Ecole* included many names from *Mankind Quarterly* including Gayre, H.E. Garrett, Luigi Gedda (who achieved brief notoriety in 1961 with a profoundly unsympathetic study of the offspring of Italian women and black American servicemen), Bertil Lundman and R. Travis Osborne. It also included Arthur Koestler, Konrad Lorenz, Robert Ardrey, H.J. Eysenck and Raymond B. Cattell.

Nonetheless the objects of the 'new' right in the USA and Britain were distinct from those in France. The right emerged everywhere as a reaction to events of the 1960s but in America the causes were different. Some American observers were aware of the connection between the political changes of the 1970s and the increasing fashionableness of previously unpopular views. One pointed out the surprising difference between the paper of Arthur Jensen in 1967 on educability and IQ and the more famous subsequent one in 1969. In the first, Jensen had stated, 'IQ is almost totally unpredictive of learning ability in the low-IQ range of low socio-economic status children.'[32] In the subsequent article which achieved notoriety in the *Harvard Educational Review* for its attack on compensatory education programmes, Jensen had said almost the opposite. The writer felt that,

> The best explanation for the divergence in views relies not on factors internal to the scientific discipline — such as new data or improved theories — but on the social and political climate within which the discipline functions. It *is* just that 1968 (a national election year) fell between the dates of the two publications . . . The election transferred political power from an administration sympathetic to these theorists who promised data and hypotheses that justified the social welfare programmes of the Great Society to an administration prepared to listen to and support the work of more pessimistic and conservative theorists and prepared to use their findings as a rationale for cutting back such social welfare programmes.[33]

As the relatively prosperous 1960s moved into the 1970s with mounting inflation and industrial unrest, similar developments could be seen in Britain. The right began to formulate three principles — a more full-blooded defence of capitalism, a demand for

The New Right

greater industrial discipline and the curtailment of welfare. Sir Keith Joseph's initial — and rather badly received — speeches on eugenic deterioration and the working-class birth rate were made at a low point for conservatism. A miners' strike had been successful and a Conservative government had lost an election called on the issue.[34] 1974 saw the beginnings in Britain, due to these events, of a 'little fear' on the right, equivalent to the 'great fear' of 1968 in France.

What role was eugenics and mental testing to play in the reformulation of right wing ideology in the politics of the 1970s? First it was as a defence of capitalism. Whereas the French right or sections of it might be squeamish about this, the British and American right clearly felt social and economic success and natural merit went hand in hand. Second, eugenics and mental testing were tools to attack the efficacy of welfarism. What good could be brought about by environmental change when one's position in the social hierarchy was ultimately determined by heredity? Like their nineteenth century counterparts, the twentieth century apologists for capitalism felt that a moral revoluton — a revolution in the hearts and minds of the populace — was necessary for its views to gain acceptance. This might paradoxically meant greater state intervention in order to enforce individual responsibility and to police the effects of the increasing hardship that a free-market economy might bring. Above all, it was necessary constantly to reiterate the natural and scientific reasons why some succeeded in these conditions and others failed.

Nonetheless occasionally the wires got crossed. A straightforward defence of capitalism does not involve anti-semitism. It was important for those on the 'new right' who were aware of the political opportunities brought by the 1970s not to alienate support by raising that spectre. For a section of the new right, anti-semitism was irrelevant anyway. Moreover, overt anti-semitism would injure the prospects of the right in Britain and America in the 1970s. Consider the historical re-evaluation of eugenics. It was no longer necessary to rehabilitate eugenics as part of a social welfare programme as Blacker had attempted in the 1940s. Welfarism was no longer as universally popular as it had been then. However, it was still vital to cut the knot binding it to Naziism. A reassessment of its historical role was aimed at reincorporating eugenics as part of 'normal practice' before the 'distortions' of Nazi practice. Thus Jean-Jacques Moreau's history of eugenics in *Nouvelle Ecole* 1971 insisted on the widescale and 'respectable' nature of eugenics in 1939.

The New Right

At the moment when sinister rumours of the Second World War were spreading, the eugenic ideal was almost universally accepted. The work of Francis Galton was known throughout the entire world. Far from being the appendage of one country, extremely concrete eugenic measures sanctioned by strong laws had been adopted by nations where very different political systems reigned. A tour of the world horizon is sufficient to convince one of this.[35]

In fact Moreau's historical knowledge was patchy. Nonetheless to assert the political universality of eugenics was a key theme in subsequent revisionist work. In particular its aim was to cut the link between the history of the extreme right and eugenics. In some cases it resulted in the rediscovery of 'left wing' eugenics. A whole literature filtered into academic life based on the exaggerated emphasis of the eugenic mavericks whose politics did not fit into the normal spectrum of pre-war eugenic thought. Their uniqueness was intended to be re-established as the norm. The object was to block out a proper historical appreciation of the social role eugenic thought actually played in pre-war society and prepare a new historical role for it in the changed circumstances of the 1970s.

Notes

1. M.F. Berry and J.W. Blassingame, *Long memory: the black experience in America* (Oxford University Press, New York, 1982).
2. Ruggles Gates Papers, RG1/9, RRG to Rev Vernon F. Gallagher (indexed circa 1948).
3. Ibid., RG1/8, memo from students and staff to Dr J. St Clair Price.
4. Ibid., RG1/9, RRG to Rev Vernon F. Gallagher, (1948), Fn 2.
5. Ibid., RG1/10, Blaiston Company to RRG (28 March 1949).
6. Ibid., RG1/10, L. Brimble to RRG (15 April 1949).
7. Ibid., RG1/9 RRG to H.E. Garrett (20 May 1948): 'I think it is up to you to defend yourself in *The Scientific Monthly.*'
8. Ibid., RG1/16 W.J. Simmons, secretary, Jackson Citizens Council to RRG (24 May 1955).
9. Ibid., RG1/17, W.J. Simmons (this time under the banner of the Association of Citizens Council) to RRG (28 June 1956).
10. Ibid., RG1/21 James A. Gregor to RRG (17 November 1960).
11. Details of these organisations can be found in Barry Mehler, 'The new eugenics academic racism in the US today', *Science for the People* (May-June 1983), pp. 18-23; and in Michael Billig, *Psychology, racism and fascism* (Searchlight Booklet, Birmingham, 1979).

12. Ruggles Gate papers H.E. Garrett to R.R. Gates (23 November 1959).

13. Ibid., RG1/19, R.R. Gayre to RRG, 8 December 1958, 'Do remember my great personal friend Prof Martino, former rector of Messina until last year, where I appointed him in 1943, and former Foreign Minister and still MP . . . If you ever want a bit of pressure to be put on in Rome, I can do it through him.'

14. Ibid., RG1/20, Gayre to RRG (28 October 1959).

15. Ibid., RG1/21, Gayre to RRG (14 March 1960).

16. James A. Gregor, 'Report on the nineteenth international congress of the Institute of Sociology in Mexico City', *Mankind Quarterly*, vol. 1, no. 2 (1960), p. 128.

17. Dr Karl Smith, 'Text of the Congressional investigation into testing', *American Psychologist*, vol. 20 (1965), p. 911.

18. Arthur H. Brayfield, Evidence to the Congressional Committee into Testing, ibid., p. 889.

19. Philip Ash, The implications of the Civil Rights Act for psychological assessment in industry, ibid., vol. 21 (1966), p. 798.

20. C.P. Blacker, *Eugenics, Galton and after*(Duckworth, London, 1952), p. 139.

21. Eugenics Society Papers, Eug C10, J.R. Baker to C.G. Bertram (14 August 1960).

22. Robert Ardrey, *The territorial imperative* (Atheneum Press, New York 1966) Konrad Lorenz, *On aggression* (Harcourt Brace and World, New York, 1966) Desmond Morris, *The naked ape* (Cape, London, 1967) A.R. Jensen, 'How much can we boost IQ and scholastic achievement?' *Harvard Educational Review*, vol. 39 (1969) C. Jencks, *Inequality* (Basic Books, New York, 1972) H.J. Eysenck, *The inequality of man* (London, Temple Smith, 1973) J.R. Baker, *Race* (Oxford University Press, Oxford, 1974) E.O. Wilson, *Sociobiology: the new synthesis* (Belknap, Cambridge, Mass, 1975) D.P. Barash, *Sociobiology and behaviour* (Elsevier, New York, 1977) R. Dawkins, *The selfish gene* (Oxford University Press, Oxford, 1976) D.L. Eckberg *Intelligence and race* (Praeger, New York, 1979)

23-30. The Baker Papers were consulted for this section.

31. See Julien Le Brunn, *La nouvelle droite* (Oswald, Paris, 1979).

32. Gerald Dworkin, 'Comment', *American Psychologist*, vol. 29, (June 1974), p. 465.

33. Ibid., p. 466-7.

34. See *The Guardian* (21 October 1974).

35. Jean-Jacques Moureau, *Nouvelle Ecole*, no. 14 in (Brunn, *La nouvelle droite*, Fn 31) (Jan-Feb 1971) pp. 126-7: 'Au moment où se font entendre les premières et sinistres rumeurs de la seconde guerre mondiale, l'idéal eugénique est presque universellement admit. L'oeuvre de Francis Galton est repandue dans le monde entier. Loin d'être l'apanage d'un pays, des mesures eugéniques extrèmement concrètes sanctionées par les legislations en vigueur ont été adoptées par des nations où règnent des systèmes politiques tres differents. Un tour d'horizon mondial suffit à s'en convaincre.'

8
Conclusion

The 1940s were a period of major structural and political upheaval in world politics. Science shared in this. In the development of atomic power, scientists entered into a close relationship with government. In the process their political and professional loyalty was put to test. It was not only that government demanded their adherence. Because scientists were in a particularly strong position to inform and direct public opinion about the nuclear issue, many entered into the political arena in the anti-atomic bomb movements of the 1950s.

However just at the moment when scientists stepped into the public arena, public confidence in science declined. In the early 1950s *The Scientific Worker*, journal of the Association of Scientific Workers, complained about the way scientists were being portrayed in film and fiction. They were depicted as Mephistopheles-type figures, tempting politicians with the secret of great power, indifferent to human suffering. Never had the power of science been so high and public confidence in it so low. This was reflected, claimed *The Scientific Worker*, in a drop, in the early 1950s in the number of British school students opting for a scientific education.

At the same time atomic scientists suffered from the attentions of government. In the late 1940s and early 1950s they found the publication of research results, dissemination of information at conferences and foreign travel increasingly difficult. They experienced a freezing of the atmosphere towards their professional collective activity. Their role as purveyors of knowledge and power was important to government on the one hand but, on the other, they were required to be subordinate and loyal employees.

This was as true in the Soviet Union as in the West. The 1950s saw the links cut between Soviet and Western science. There was

Conclusion

a drive against foreign contacts and cosmopolitanism which had aided the triumph of Lysenkoism in Soviet biology. This triumph had the effect of strengthening the divisions between Western scientists which arose in the 1940s and which revolved around the issues of socialism, atomic power and the deteriorating international climate. The Soviet Union was offering not only an alternative social system and means of planning scientific development but also an 'alternative' science at least as far as biology was concerned.

The mephistophelean view of science persisted throughout the 1950s and was fuelled by the atomic weapons test programmes. Public insecurity was in fact justified. The bland reassurances of government covered a great deal of ignorance and 'feeling the way' in the atomic tests of this epoch. When the extent of nuclear fall-out was revealed in the *Lucky Dragon* episode of 1954, governments shared public apprehension. They had, in fact, embarked upon a course of action when they had only the most hazy understanding of the ultimate consequences. Meanwhile the secrecy around these tests had been self-defeating because, in the end, it served only to increase public alarm.

Out of this situation in the 1950s came two developments. One was serious research into the effect of fall-out and of a world in which nuclear energy was going to play an increasingly important part. The other was the Pugwash movement which helped to recreate some form of international scientific co-operation. By the late 1950s the fact that public alarm about testing was increasing, and the existence of Pugwash helped pressurise the atomic powers into the political initiatives which led eventually to the Test Ban Treaty of 1963.

If the 1940s had brought division among scientists and increasing political control over science it had also more positive effects. In the 1930s Professor Zollschan had tried and failed to get international action on Nazi racial doctrine. It took the impact of the Second World War and the revelations about Nazi eugenics before any effort in this direction was successful. The creation of UNESCO, with the aim, among others, of European re-education gave an opportunity to pre-war anti-Nazis to unite the fundamentals of anti-racism into a major document, the UNESCO statement on race. Discussion on the drawing up of this revealed how incomplete was the conversion of many anthropologists and geneticists to anti-racism. Nonetheless, the UNESCO document did have an effect.

In the 1950s, however, the Cold War in Europe and America led to other ideological imperatives: in particular the reformulation of Western ideals to counter Communism. In the case of science, this led to an attack on the influence of Bernalism and the reformulation of science to fit in with the way in which protagonists of the West conceived their society. In the view of Polanyi and others, science was the supreme example of capitalism at work. But the capitalism it reflected was the nineteenth century view of the free competitive market in which there was opportunity for all to enter and the best survived. Thus Polanyi contrasted the view of scientific ideas a products tested in the intellectual market with the Bernalist view of state-managed planned science. Similarly Edward Shils, in his newspaper articles of the late 1940s before the Soviets got the bomb, pictured Soviet scientific backwardness as the product of the country's semi-oriental system of government which had escaped the process of capitalist evolution. Much of the sociology of science during this period emphasised the scientist as an entrepreneur and his product, scientific discovery, as the means of social and economic advancement.

This view ignored the deep commitment of governments, East and West, in the 1940s and 1950s to investment in and control of science. It also ignored the employee status and subordinate position of most scientists. Thus in reality both scientific trades unionism persisted and so did many of the professional frustrations of scientists at this employee status, at the inadequate provision of support for science (other than defence) and at the restrictions which government and industry posed. The successful entrepreneurial scientist was certainly a role model for some but even Cherwell, who might claim to have emulated this model more than most, realised that the status and role of science was much more fragile. By the late 1950s when Sputnik was launched, the confidence of those who believed that the free market would ensure continued Western superiority was shaken even further. In the USA, a concerted effect at government investment in scientific education and research — a kind of Bernalism in practice — was the result.

In Britain too, Bernal's ideas, in spite of the ideological changes of the 1950s, continued to have influence, largely because modern science was not just a reflection of free-market activity. In the 1950s Bernal was, along with Blackett, a member of a discussion group on science organised by Marcus Brumwell, to which many members of the Labour Party's shadow cabinet were invited. It was here that planning and an increase in scientific education were discussed. There was, in these discussions, resistance on the part of Hugh Gaitskell who, although he had been an officer of the AScW and was

Conclusion

sympathetic to scientists, was deeply suspicious of the Bernalist belief in large-scale planning. Gaitskell was unwilling to support this at the same time as he was attempting to remove the commitment to state control of industry in the Labour Party programme. Moreover he was convinced of the ability of capitalism to produce wealth, if not to distribute it fairly.

However, Harold Wilson was, in his government of 1964-6, to take on many of Bernal's ideas. A Ministry of Technology was created in which Blackett served and which helped create the British computer industry. Interestingly Blackett concluded from his experience that small-scale enterprises were sometimes better mediums for long-term planning. On the whole, however, this venture of a Ministry of Technology was an indirect tribute to Bernal's influence. So too was the educational expansion seen in these years.

But if the Polanyi view of science as capitalism was ineffective when faced with the problems of scientific research and development, it none the less served as a means of re-education. The idea that science reflected the values of private enterprise proliferated in journals, newspapers, learned books and intellectual conferences. But another less friendly re-evaluation of science was taking place at the same time. Polanyi was a believer in reason and progress. But for others if science had served to produce the atomic bomb and the Nazi racial experiments, then this put in doubt the intellectual foundations of science itself. In an exercise which neatly side-stepped the issue of politics and the state, they sought an explanation in the philosophy of science. Out of this came the stirrings in the late 1950s of an attack on the ethos of science and a search for 'better' and 'sounder' values in literature or ethical philosophy. This attack associated Communism with all the attitudes it ascribed to science — belief in human manipulation, progress for its own sake, destruction of the traditional, reason valued above love or devotion. In this view, scientism and Communism were two aberrations which naturally followed upon one another. For others, the criticism of the dehumanisation produced by science extended to a critique of the foundations of modern society itself. For these people, capitalism as well as Communism fell under suspicion. Both certainly used and valued science.

In the 1960s both these views reached their fullest expression in the anti-mental testing movement and the search for social systems alternative to those offered by both East and West. Whilst this period of intellectual as well as social upheaval was brief, it

Conclusion

stimulated a re-examination of many areas of activity. In addition, it produced greater consciousness of when the social system was using the name of science to propagate its own values. It also caused some scientists to be more aware of the temptation of slipping into the role of spokesman or woman for the social order.

Bibliography

G.A. Almond (ed.) *The appeals of Communism* (Princeton University Press, Princeton, 1954)

Lawrence Badash, J.O. Hirschfelder and P. Broida (eds) *Reminiscences of Los Alamos, 1943-5*, (Kluwer, Boston, 1980)

J.R. Baker *Race* (Oxford University Press, Oxford, 1970)

C.J. Bartlett *The long retreat: a short history of British defence policy 1945-70* (Macmillan, London, 1972)

E. Baur, E. Fischer and F. Lenz *Human heredity*, 3rd edition, 1927; translated and published in English (George Allen and Unwin, London, 1931)

J.D. Bernal *The social function of science* (Routledge and Kegan Paul, London, 1939)

—— *Science in history* (Watts and Co, London, 1954)

M.F. Berry and J.W. Blassingame *Long memory: the black experience in America*, (Oxford University Press, New York, 1982)

M. Billig *Psychology, racism and fascism* (Searchlight, Birmingham 1977)

Earl of Birkenhead *The Prof. in two worlds* (Collins, London, 1961)

C.P. Blacker *Eugenics, Galton and after* (Duckworth, London, 1952)

Patrick M.S. Blackett *The military and political consequences of atomic energy* (Turnstile Press, London, 1948)

Julien le Brunn *La nouvelle droite* (Nouvelles Editions, Oswald, Paris, 1979)

Elof Axel Carlson *Genes, radiation and society: the life and work of H.J. Muller* (Cornell University Press, Ithaca, NY, 1981)

Linda Clark *Social Darwinism in France* (University of Alabama Press, 1985)

Ronald W. Clark *J.B.S.: the life and work of J.B.S. Haldane*, (Hodder and Stoughton, London, 1968)

—— *Life of Bertrand Russell* (Cape, London, 1975)

—— *Tizard* (Methuen, Londo, 1965)

John Langdon Davies *Russia puts the clock back* (Gollancz, London, 1949)

N. Deakin *et al The new right: image and reality* (Runneymede Trust, London, 1986)

C. Driver *The disarmers, a study in protest* (Hodder and Stoughton, London, 1964)

K. Dronamraju (ed.) *Haldane and modern biology* (John Hopkins, Baltimore, 1968)

J. Fyfe *Lysenko is right* (Lawrence and Wishart, 1950)

Robert Gilpin *American scientists and nuclear weapons policy* (Princeton University Press, 1962)

M. Goldsmith *Sage: a life of J.D. Bernal* (Hutchinson, London, 1980)

M. Goldsmith and A. Mackay (eds) *The science of science* (Souvenir Press, New York, 1964)

G. Goodman *Awkward warrior: a life of Frank Cousins* (Davis-Poynter, London, 1979)

M. Gowing *Independence and deterrence: Britain and atomic energy*, 2 vols (Macmillan, London, 1974)

A.J.R. Groom *British thinking about nuclear weapons* (F. Pinter, London, 1974)

Bibliography

A.C. Haddon and Julian Huxley *We Europeans* (Cape, London, 1935)
Otto Hahn *My Life* (Macdonald, London, 1970)
John Hendry (ed) *Cambridge physics in the thirties* Adam Hilger, Bristol, 1984)
Greta Jones *Social Darwinism and English thought* (Harvester, Brighton, 1980)
—— *Social hygiene in twentieth century Britain* (Croom Helm, London, 1986)
D. Joravsky *The Lysenko affair* (Harvard University Press, Cambridge, Mass, 1970)
Daniel K. Kevles *In the name of eugenics, genetics and the uses of human heredity* (A. Knopf, New York, 1985)
Dominique Lecourt *Proletarian science?* (New Left Books, London, 1977)
Robert Jay Lifton *The Nazi doctors, medical killing and the psychology of genocide* (Macmillan, London, 1986)
Kenneth Ludmerer *Genetics and American society* (John Hopkins, Baltimore, 1972)
Alwyn MacKay *The making of the atomic age* (Oxford University Press, Oxford, 1984)
Z.A. Medvedev *The rise and fall of T.D. Lysenko* (Columbia University Press, New York, 1969)
E. Meehan *The British left wing and foreign policy* (Rutgers University Press, New Brunswick, 1960)
Robert Milliken *No conceivable injury: the story of Britain and Australia's atomic cover-up* (Penguin, Harmondsworth, 1986)
Walter Millis (ed.) *The Forrestal diaries* (Cassell and Co, London, 1952)
Alan Morton *Soviet genetics* (Lawrence and Wishart, London, 1951)
Norman Moss *Klaus Fuchs, the man who stole the atom bomb* (Grafton Books, London, 1987)
Frank Parkin *Middle class radicalism: the social bases of the Campaign for Nuclear Disarmament* (Manchester University Press, Manchester, 1968)
Henry Pelling *The British Communist Party* (Adam and Charles Black, London, 1958)
R. Peierls *Bird of passage: recollections of a physicist* (Princeton University Press, New Jersey, 1985)
R.M. Pike 'The growth of scientific institutions and the employment of natural science graduates in Britain 1900–60, MSc. University of London, 1961
R. Ruggles Gates *Pedigree Negro families* (Blakiston Press, Philadelphia, 1949)
H. & S. Rose (eds) *The radicalisation of science* (Macmillan, London, 1976)
G. Searle *Eugenics and politics in Britain 1900–14* (Noordhof, London, 1976)
Nancy Stepan *The idea of race in science, Great Britain* (Macmillan, London, 1982)
A.H. Teich *Scientists and public affairs* (Harvard University Press, Cambridge, Mass, 1973)
UNESCO *The race concept: results of an inquiry* (UNESCO, Paris, 1952)
Gary Werskey *The visible college* (Allen Lane, London, 1978)
Philip M. Williams *Hugh Gaitskell* (Cape, London, 1979)
—— (ed) *The Diary of Hugh Gaitskell, 1945–56* (Cape, London, 1983)
C. Zirkle *Evolution, Marxian biology and the social scene* (University of Pennsylvannia Press, Philadelphia, 1949)
Ivan Zollschan *The significance of the racial factor as a basis in cultural development* (Le Play House, London, 1934)

Bibliography

Manuscripts

Association of Scientific Workers Papers, Modern Records Centre, University of Warwick Library
Baker Papers, Bodleian Library, University of Oxford
Blackett Papers, Library of the Royal Society, London
Burhop Papers, University College London, Department of Physics
Burt Papers, University of Liverpool
Cherwell Papers, Nuffield College, Oxford
Eugenics Society Papers, Wellcome Unit for the History of Medicine, London
Peierls Papers, Bodleian Library, University of Oxford
Penrose Papers, University College, London
Ruggles Gates Papers, Kings College, London
Society for Freedom in Science, occasional papers, Oxford (1945)
World Federation of Scientific Workers Papers, Modern Records Centre, University of Warwick Library

Government Papers and Publications

Cabinet Papers
CAB 128/4
CAB 128/27
CAB 129/4
CAB 130/8
Manual of basic training, Civil Defence II (HMSO, London, 1950)
Cmnd 9780 The hazards to man of nuclear and allied radiation (HMSO, London, 1956)
Cmnd 363, Report on Britain's contribution to peace and security (HMSO, London, 1958)
Cmnd 1225, The hazards to man of nuclear and allied radiation Medical Research Council, Cmnd 9780, (HMSO, London, 1960)

Other Printed Sources

Reports of the Proceedings of the Trades Union Congress

Newspapers

Coventry Evening Telegraph
Daily Herald
Daily Worker
Evening Standard
Newcastle Journal
News Chronicle

Bibliography

Manchester Co-operative News
Manchester Guardian (later *The Guardian*)
Sheffield Telegraph
South Yorkshire and Rotherham Advertiser
The Times

Periodicals

American Anthropologist
American Historical Review
American Journal of Physical Anthropology
American Psychologist
Annals of Science
Archives of Psychology
Atomic Scientists News (later *Atomic Scientists Journal*)
British Medical Journal
Bulletin of Atomic Scientists
Discovery
Eugenics Review
International Social Science Bulletin (UNESCO)
Isis
Journal of Heredity
Journal of Modern History
Labour Monthly
The Listener
Man
Mankind Quarterly
Minerva
Modern Quarterly
Nature
New Statesman
Nineteenth Century and After
Political Science Quarterly
The Realist
Reconstruction and Rehabilitation Newsletter (UNESCO)
Science
Science and Society
Science for the People
Scientific Monthly
The Scientific Worker

Index

Acheson — Lilienthal Plan 82
Acland, Sir Richard 91
Amalgamated Engineering
    Union (AEU) 103
American Anthropological
    Association 59
American Association of Atomic
    Scientists 108
American Psychologists
    Association 126
Anderson, Sir John 13, 92-3
anti-Semitism 5, 60, 63-6, 75,
    120, 129-30, 133
Appleton, Sir Edward 39
Ardrey, Robert 132
Arnott, R. 30
Ashby, Eric 17-18, 22, 23, 25,
    26
Association of Scientific
    Workers (AScW) 12-14, 93,
    98, 111
    membership 8, 50
    views on atomic Arms Race
        87-9
    views on Lysenko 39-56
Atomic Bomb Casualty
    Commission 99
Atomic Energy Bill (Britain) 42
Atomic Energy Commission of
    the United Nations (AEC)
    81, 96
Atomic Scientists Association
    39-56, 79, 86, 108
Atomic Weapons 79-94
    and civil defence 92-3,
        101-3, 111-12
    and middle opinion 90-2
    and military strategy 84-6
    and the ASA 86
    and the AScW 87-9
    and the TUC 89-90
    British atomic bomb 81-3
    international control of 80-1
Atoms for Peace Conference
    105, 117n34

Attlee, Clement 80, 82, 104
Auerbach, Charlotte 107
Auschwitz 59-60, 66
Australian Royal Commission
    on Atomic Tests 99

Baker, J.R. 12
    and Lysenko 28-9
    and the SFS 21-3
    on UNESCO 70-2
    relations with the 'new right'
        128-30
Baruch Plan 81, 82, 85, 87, 111
Bateman, Angus 48
Baur, Erwin 63-4, 74, 131
Beaglehole, Ernest 67
Belsen 59
Benes 1
Benoist, Alain de 129, 131
Bernal, J.D. 9, 25, 53, 55, 93
    and Lysenko 31-4
    and on Labour Party 138-9
    influence on Pugwash 108
    views on science and
        capitalism 11-14
Bertram, C.G. 128
Bevin, Ernest 8, 17, 39
Biquard, Pierre 108
Blackburn, Raymond 90
Blacker, C.P. 28, 65, 70, 128-9
Blackett, P.M.S. 8, 12, 46
    and the AScW 13-14
    and the Ministry of
        Technology 139
    and the TUC 89
    on atomic strategy 39, 83-8
    on civil defence 92-3
Blakeslee 51
Blakiston Press 120
Boas, Franz 74-5
Boder, Jaime Torres 66
Bohr, Nils 4, 5
Born, Max 43, 93, 109
Botanical Congress Stockholm,
    1950 51

Index

Brabazon, Lord 90
Brayfield, Arthur H. 126
Bridgman, P.W. 55
Brimble, L. 120-1
British Association for the Advancement of Science 2, 12, 22, 24
British Broadcasting Corporation (BBC) 24, 25-6, 28, 102-3
British Industry 6-7
Brittain, Vera 91
Brumwell, Marcus 138
Buchenwald 60
Bukharin, N. 11
Burhop, E.S. 45-7, 79, 86, 108, 110
Burt, Cyril 76-7

Calder, Ritchie 28
Cambridge Scientists Anti-War Group 13
Campaign Against Nuclear Tests 104
Campaign for Nuclear Disarmament (CND) 90, 104-5, 114, 116
Canaris, Admiral 60
Cantril, Professor Hadley 67
Carnegie Institute 1
Cattell, Raymond B. 132
Cavendish Laboratory, Cambridge 4-6, 12, 83
Chadwick, Sir James 39
Champion, J. 47
Chemical Workers Union 89, 98
Cherwell, Lord (Lindemann R.) 14, 51, 84, 138
    and the ASA 46-7
    on fall out and civil defence 101-4
Church of England Commission 80
Churchill, Winston 101-4
Civil Defence see Atomic Weapons
Civil Rights Act (USA), 1964 127
Civil Service 24, 46
Clark, Professor Le Gros 69

Comas, Juan 67, 125
Committee on Science and Freedom 52, 57n44, 110
Committee to Bring Morality to the Mental Professions 126
Communist Party of Great Britain 11, 30-1
Conference of Parliamentarians for World Government 109
Congress for Cultural Freedom 72, 110
Conklin, E.G. 68
Conti, Dr 66
Cook, Arthur 10
Co-operative Party 111
Cousins, Frank 115
Coventry City Council 103, 111
Crew, F.A.E. 3
Cripps, Sir Stafford 80
Crossman, R.H.S. 111-2
Cudlipp, Hugh 43
Cummings, A.J. 27-8

Dachau 60
Dahlberg, Gunnar 68
Dale, Sir Henry 26-8, 39, 45, 52, 54
Darlington, C.D. 23-4, 26, 30, 51, 70, 124, 128
Darwin, Charles 32, 125
Darwin, Sir Charles 128
Davenport, C.B. 75
Davies, Clement 90
Davies John Langdon 28
Davies, R.G. 19
Dawkins, R. 131
Deakin, Arthur 115
Defence White Paper, 1955 112
    Defence White Paper, 1957 93, 111, 113-4
    Defence White Paper, 1958 114
Dobzhansky, T. 17, 68
Doll, Dr Richard 91, 107
Draper, Wycliffe 123
Drzewieski, Dr 67
Dunn, L.C. 68
Dutt, Clemens 30

Eastland, James O. 123

146

Index

Eaton, Cyrus 109
Eickstedt, E. von 124
Einstein, Albert 109
Elements, 131
Eliot, T.S. 90
Ellinger, H. 119-20
Emery, Peter 103
Equal Opportunities Act (USA), 1972 127
Ervin, Sam 125
Eugenics 4, 10, 13, 28
  and the 'new right' 122, 128, 133-4
  and UNESCO document on race 70-1
  German eugenics 62-6
  interwar eugenics 60-2
Eysenck, Hans 129, 132

Fabian Society 88, 91
Fall Out *see* radioactive fall out
Federation of American Scientists 38
Ferri, Enrico 4-5
Fisher, Eugen 63-4, 71, 74, 124, 131
Fisher, R.A. 19, 24, 26, 30, 70, 124
Fleure, H.J. 69
Forrestal, James 80
Foundation for Human Understanding 122
Frankfurt Institute For Hereditary Biology and Racial Hygiene 64
Frazier, Franklin 67
Freitas, Geoffrey de 103
Frich, Dr Hils 66
Frick, Wilhelm 65-6, 72
Frisch, Otto 6, 40, 43
Frisch — Peierls Memorandum 6, 40
Fuchs, Klaus 40, 42-4
Fyfe, J.L. 19-20

Gaitskell, Hugh 112-13, 138-9
Garrett, H.E. 121-3, 127, 132
Gates, R.R. 1, 74, 119-22, 127
Gayre, R.R. 123-4, 129, 132

Gedda, Luigi 132
General Council of the Trades Union Congress (TUC) *see* Trades Union Congress
Genetical Society 105
Genetics 16-23
  and Marxism 31-4
  and race 72-4
  passim 62-72
Geneva Convention 60
Gerlach, Walter 110
Gini, Corrado 122, 124
Ginsberg, Morris 67
Gray, J.L. 75
Gregor, James A. 122, 125
Gregory, Sir Richard 10
Groom, A.J. 111
Groupement de recherche et d'etude pour la civilisation europeene (Grece) 131
Gustafsson, Ake 51

Haddon, A.C. 75
Hahn, Otto 4, 109, 110
Haldane, Charlotte 28
Haldane J.B.S. 3, 12-13, 49, 75, 100, 128
  and Lysenko 18-21, 25, 27-34
Harland, S.C. 3, 24, 26
Harwell Atomic Weapons Establishment 98, 100, 107
Herskovits, M. 74
Hessen, B. 11
Hill, Dr Osman 69
Hill, Raymond E. 124
Hinden, Rita 91
Hiroshima 97, 99
Hirt, Professor 60
Hogben, L.T. 3, 75, 128
Howard University 119-20
Human Betterment Foundation of the USA 65
Human Genetics Fund of New York 123
Huxley, J.S. 3, 10. 75
  and Lysenko 18, 22, 34
  and UNESCO 66, 68
hydrogen bomb 46, 82, 96, 98-9, 100

Index

Imperical Chemical Industries 9-10
Imperial Defence College 85
Institut International de Sociologie 122
Institute for the Study of Education and Differences 122
Institute for the Study of Man 122
Institute of Industrial Psychology 10
Institute of Intellectual Co-Operation 1
Intelligence Testing 69-77, 25-*passim* 1
International Association for the Advancement of Ethnology and Eugenics 122
International Committee of the Genetics Congress 3
International Confederation of Free Trades Unions (ICFTU) 89, 115
International Conference of Pure and Applied Chemistry, 1947 79
International Congress of the Anthropological Sciences 1
International Radiological Protection Board 98, 106

Jackson's Citizen's Council 121
Jensen, Arthur 132
Joffe, A.F. 17
Joliot-Curies 4, 6
Joliot-Curie, Frederic 6, 9, 83, 107-9
Joravsky, D. 17-24
Joseph, Sir Keith 133
Joule, Dr H. 91

Kabir, Dr Humayin 67
Kaiser Wilhelm Institute, Berlin 4
Kapitza, P.L. 4, 7
Keith, Sir Arthur 69
Kissinger, Henry 85
Klineberg, Otto 68, 74, 75
Koestler, Arthur 79, 132

Kurti, N. 37, 43

Labour Party 49, 104-5, 112-13, 115, 138-9
Lamarck 125, *see also* Lysenko
Lang, Reverend Gordon
Langevin, Paul 83
Laue, Max von 110
Lenin Academy of Agricultural Sciences *see* Soviet Union
Lenz, F. 63-4, 71, 74, 124, 131
Levi Strauss, Claude 67
Levit, S.G. 3
Levy, Hyman 12, 30
Lewis, John 19-20
Lindau Appeal 109
Lonsdale, Kathleen 47, 91, 93, 108
Lorenz, Konrad 131-2
Loughlin, Harry 123
Loutit, J.F. 100
Lucky Dragon 96-7, 103-5, 108, 115
Lundman, Bertil 132
Lysenko 3, 38, 69, 71, 128, 137
  and AScW 48-51
  controversy over 16-35
  effect on ASA 45-6, 86

Mackintosh, Professor J.M. 91
MacMillan, Harold 107
Manhattan Project, Los Alamos 6, 39, 40
Mankind Quarterly 123-5, 128, 132
Marley, Dr W.G. 99
Marshall Aid 82, 89
Martin, Kingsley 83
Martino, Professor 123
Marxim Gorky Medico Genetics Institute 3
Marxism 11, 16-35 *passim*
Massey, Harrie 47, 108
Medical Association for the Prevention of War (MAPW) 91-2, 105
Medical Research Council 98, 100, 105-7
Meitner, Lise 4, 5
Mendelism 16-19, 21, 26, 32-3
  see also genetics

Index

Mengele, J. 66
Metraux, A. 2, 69
Michurinism 16, 21
Ministry of Technology 139
Minnesota Psychological Tests 125-6
Molotov 41
Mond, Alfred 9-11
Mondism 9-11
Montagu, Ashley 67, 74-5
Moore, Wilbert 68
Moran, Lord 65
Morant, D. 69
Moreau, Jean-Jacques 133-4
Morganism 22-33 passim
Morrison, Herbert 26-7, 51
Morton, Alan 33
Mott, Nevil 44, 47, 108
Mountbatten, Lord 13, 86
Muller, H.J. 3, 38, 62
  and radioactive fall out 98, 105, 107
  and UNESCO 68, 69-70
Myart, Leon 126-7
Myrdal, Gunnar 68

Nachtsheim, Hans 71-2
Nagasaki 97, 99
National Peace Council 91
National Union of Scientific Workers (NUSW) 7-8 see also Association of Scientific Workers
Nazis 1, 2, 4, 5
  and eugenics 62-6
  de-nazification 71, 78n24, 137
  experiments on human subjects 59-60
Needham, J. 3, 68
New Right, The 130-4
Nietzsche 64, 131
Nixon, W.G.W. 91
Non-Conformist Churches 104
Noontide Press 122, 129
North Atlantic Treaty Organisation (NATO) 82, 85
Nouvelle Ecole 131-3
Nunn May, Alan 42-3
Nuremburg Trials 60, 65

O'Brien, Tom 90
Oparin, A. 108
Oppenheimer, Philip 82
Oppenheimer, Robert J. 99
Orr, John Boyd 91, 93
Osborn, Frederick 123
Osborne, R. Travis 132

Padley, W. 90
Peace Corps 125-6
Peace Pledge Union 91
Peierls, Rudolf 6, 79, 86
  and the ASA 39-41, 46-8
  and Klaus Fuchs 43-4
Penrose, L.S. 69, 73, 91, 100, 105
Pinto, L.A. Costa 67
Pioneer Fund 122-3
Pirie, N. 48
Polanyi, M. 12, 22, 34, 79, 129
  views on the structure of science 53-5, 138, 139
Polyakov, I.M. 19
Pontecorvo, Bruno 42-3
Popenoe, Paul 64-5
Powell, C.F. 89, 109
Prezent, I.I. 16
Progressive League 10
Pugwash 109-11, 137
Putnam, Carleton 129-30

Quakers 91, 115
Quibell, Lord 90

Rabinowitch, Eugene 108-9
racism 1-4
  and genetics 68, 72-4
  and intelligence 69-77 *passim*, 125-7
  in the United States 119-22
  see also anti-semitism
radioactive fall out 96-107
Ramos, Dr A. 67
Rasche, Major 60
Robertson, Wilmot 129-31
Robinson, Sir Robert 26-7, 39
Rosenbergs 42
Rotblat, Joseph 43, 108
Rotherham Labour Party 103
Rothertham Trades Council 103

Index

Rotherham United Nations Association 104
Royal Instituttion 22
Royal Society 24, 26, 47, 51
Russell, Bertrand 90, 95n24, 109-10
Russell-Einstein Statement, 1955 109
Rutherford, Ernest 4, 5

St Clair Price, Dr J.
Saller, Karl Feliz 72
Scheidt, Professor W. 71, 124
Schilling, Karl Klaus 60
Science for Peace 93-4
Scientific Advisory Committee 83-4
Scottish TUC 111
Seligmann, C.E. 1
Shaw, G.B.S. 20, 61
Shils, Edward 39, 138
Simmons, W.J. 121
Simon 43
Skinner, H.W. 44-5, 47
Society for Cultural Relations with the USSR 53
Society for Free German Culture 53, 79
Society for Freedom in Science (SFS) 12, 17-18, 21-3, 34, 38, 52
Soukatchov, V.N. 51
Soviet Union 2, 3, 11, 14, 22
   Academy of Sciences 6, 24-6
   Institute of Genetics 16
   Lenin Academy of Agricultural Sciences 16
Sputnik 114, 138
sterilisation 60-1
Stern, Curt 68
Strassman, Fritz 4, 5
Strauss, Lewis 96
Stross, Dr B. 91
Supreme Court Decision on Segregation, 1954 121
Swann, Donald 122
Szilard, Leo 5

Tansley, A.G. 17
Teller, Edward 112
Termer, Dr Franz 59

Test Ban Treaty, 1963 110
Tewson, Sir Vincent 90
Thomson, Sir George 39, 47
Tildesley, Myres 69
Tizard, Sir Henry 39, 84
Trades Union Congress (TUC) and the Cold War 89-90, 93
   attitude to nuclear weapons 111, 113, 115-6
   drive against communists 24, 49
Transport and General Workers Union (TGWU) 8, 115
Trevor, Dr J.C. 69
Trivers, R.L. 131
Trotter, William 92

United Nations 41, 67, 88
United Nations Educational, Scientific and Cultural Organisation (UNESCO) 2, 66-72, 105, 119, 125
United States National Academy of Sciences 105-6

Van Kleffans 108
Vansittart, Lord 90
Vavilov, N.I. 20, 23, 26, 28, 45
Verschauer, Otmar von 64, 66
Vial, Pierre 131
Vyshinsky 41

Waddington, C.H. 34
Walter, Francis E. 123
Weinert, Hans 72
Werskey, G. 12
West, Rebecca 43
Wilson, E.O. 129-30, 131
Wilson, Harold 139
Windscale 107
World Federation of Scientific Workers (WFSW) 8-9, 93, 107-9
World Federation of Trades Unions (WFTU) 89, 93
Wright, Sewall 19
Wroclaw Conference 25

Zollschan, I. 1, 2, 66, 76, 137
Zuckerman, Professor S. 69